Jorge Luis Nuñez
Mario Enrique Muñoz-Cobeñas
Horacio Luis Moltedo

Boophilus microplus

The Common Cattle Tick

Translated by Harry Bailie, Constanza Michelsohn
and Cecilia Paris

With 92 Figures

Springer-Verlag
Berlin Heidelberg New York Tokyo

Dr. JORGE LUIS NUÑEZ, Cooper Argentina SACel. L.N. Alem 619,
RA-C.P. 1001 Buenos Aires

Dr. MARIO ENRIQUE MUÑOZ-COBEÑAS, Veterinary Surgeon,
University of La Plata

Dr. HORACIO LUIS MOLTEDO, Aviador Sánchez 651, Palomar,
Buenos Aires

Translators:
Dr. HARRY BAILIE, CONSTANZA MICHELSOHN, CECILIA PARIS

Title of the Argentine Edition: JORGE L. NUÑEZ, MARIO E. MUÑOZ-COBEÑAS, HORACIO L.
MOLTEDO, Boophilus microplus, © 1982 by Editorial Hemisferio Sur S. A. ISBN 950-004-239-8

ISBN 3-540-15146-X Springer-Verlag Berlin Heidelberg New York Tokyo
ISBN 0-387-15146-X Springer-Verlag New York Heidelberg Berlin Tokyo

Library of Congress Cataloging in Publication Data. Núñez, Jorge Luis, 1937–   . Boophilus
microplus. Includes bibliographies and index. 1. Boophilus microplus. 2. Boophilus microplus–
Control. I. Muñoz-Cobeñas, Mario Enrique. II. Moltedo, Horacio Luis, 1948–   . III. Title.
SF967.B64N86   1985   636.2′0894433   85-2790

Typesetting: K. u. V. Fotosatz Beerfelden
Offsetprinting and binding: Konrad Triltsch, Graphischer Betrieb, Würzburg
2131/3130-543210

Prologue

According to a publication by The Wellcome Foundation in 1978, *Boophilus microplus*, the common cattle tick, is spread over practically four continents, between parallels 30° latitude North and 48° latitude South: America from Mexico to Argentina; most of Africa, Australia, and large areas in Asia. The following questions arise: how many million head of cattle are exposed to its parasitism? What astronomical economic losses does it cause in the world of cattle exploitation?

In Argentina, for example, among cattle ectoparasitosis, *Boophilus microplus* lies in first place, and its importance, far from decreasing, is gradually increasing as the government campaign faces new difficulties when the limits of the campaign area reach the subtropical regions of the country; hence the importance of knowledge of *Boophilus microplus* and its incidence as the principal parasitosis in many cattle-rearing countries of the World.

The Argentinian publishing house Editorial Hemisferio Sur understood it as such, and felt the need to add to their long list of animal and agricultural publications, a book related exclusively to the cattle tick, on the insistent of government and private sectors of various countries interested in the problem.

Only one high-level team (made up of Drs. Jorge L. Nuñez, Mario E. Muñoz-Cobeñas and Horacio L. Moltedo) was equipped to produce a book such as that required. Happily, with foresight and wisdom, the publishers understood that this was a well-functioning team and so requested them to undertake the task of writing the book.

The authors have used an enormous amount of information available on different aspects of the ixodid, from the most ancient, such as the meticulous trials by Lahille at the beginning of this century, to the latest contributions by Australian authors.

All this enormous and complete bibliography has been consulted and subjected to critical revision to extract the most representative data and conform to the general context of the book.

The bibliography represents a work in itself, and constitutes the scientific basis of the book. Some 500 selected bibliographic references, corresponding to each of the chapters, thus provide a source of information which the authors have used rationally.

To this, we must add the world-wide personal or epistolary contact with acarus researchers of prestige and renown, who have allowed us access to valuable unpublished information and have revised some polemical points in the text.

The most important aspect, however, is the fact that the team have included their own personal experience in the text.

This experience can only result from years of working with the tick in its different aspects, in both fieldwork and in specialized laboratories.

A fundamental contribution are the morphological studies carried out by the authors with the aid of the scanning electronic microscope. It is evident that the use of this highly developed system has introduced, in the specialized centres of the world, the beginning of an era when the external morphology of parasites is being revised. The arthropods, especially due to their chitinous coverage, are very suitable for this process, and we do not doubt that the stereoscopic vision at high magnitude, of the anatomy of these creatures will allow interpretations of the morphology and behaviour of the parasites, which otherwise remain incromprehensible.

The authors have incorporated in the text magnificent illustrations obtained through SEM. Like a morphology atlas, they give us an insight into what texts on parasitology will be like in the year 2000, when the parasites will be re-examined by means of the incredible vision obtained with the scanning and transmission electronic microscope.

Over and above all the material elements (bibliography, technique, personal experience) we appreciate in this book a critical spirit, a deep conceptual honesty and the professional clarity with which the topics have been developed.

It thus becomes the most complete text on one parasite ever written in Argentina. We have no doubt that, in all Latin America where *Boophilus microplus* is endemic, it will receive a favourable reception, and wide circulation in specialized centres for animal health and veterinary science.

I here express my gratitude to the authors for requesting me to write the prologue to this fundamental work in the field of acarology, which is a tribute to the Argentine parasitology.

Corrientes   OSCAR J. LOMBARDERO

Acknowledgements

The object of these lines is to give thanks for the collaboration we have received from numerous colleagues and institutions during the 2 years (1980 – 81) we dedicated to writing this book.

To Drs. A. Pérez Arrieta, R. J. Rovere, A. Bolondi and J. C. Ivancovich of INTA; to Dr. J. Caracostantógolo of SELAB; to personnel of the Scanning Electronic Microscope Service (CONICET) and to Dr. O. J. Lombardero, professor of Parasitology and Parasitic Diseases of the School of Agronomy and Veterinary Sciences of the Universidad Nacional del Nordeste.

From abroad, we must mention Drs. H. Hoogstraal, R. A. Bram and R. Smith of the USA; Dr. R. G. Wilson of the Wellcome Research Laboratories, Great Britain; from Australia, The Australian Entomological Society and Drs. R. H. Wharton, W. J. Roulston and B. F. Stone; Dr. S. M. Waladde of ICIPE, Kenya, all of whom have collaborated by sending important information and the latest material related to the subject we are handling.

Finally, we consider it our duty to acknowledge with thanks the quantity of excellent information that for years The Wellcome Research Laboratories have provided for us, from which we have selected most of the bibliography given in this book.

The Authors

Contents

I Introduction

A detailed tracing, from acceptable sources, of archaeological and paleontological discoveries made up to the present time leads us to suppose that approximately in 8000 y B.C., in Southern Turkestan, man succeeded in domesticating the first cattle, which he later took with him as he migrated from this remote region of Central Asia.

Step by step, Europe and Asia have been gradually inhabited by domesticated cattle which have been incorporated into man's economy, both as a source of food and work.

The same happened in America and Australasia, continents where cattle were taken by the European colonizing groups during the course of the 16th to the 18th centuries. Possibly the common cattle tick also reached these continents at the same time, accompanying its most frequent host.

The cattle tick, *Boophilus microplus,* parasitizes Asiatic cattle races (with special reference to the zebu, *Bos indicus*), but generally the level of infestation is not high, only a few engorged females being detected, generally no more than ten. When cattle of European races are infested by *Boophilus microplus,* however, the level of parasitism is higher, sometimes reaching limits incompatible with the life of the host.

The main damage caused by the common cattle tick can be studied according to the following aspects:

1 Debilitating Action Exercised by a Hematophagus Parasite on a Host

Obviously the impact of the debilitating action is directly related to the number of parasites which in moderate cases reduce the weight gain of affected cattle, and in heavy infestations can cause important losses in a herd.

For example, an engorged female can suck between 0.5 and 3 ml of blood during its parasitic cycle (Barnett 1961).

As it is common to find hundreds or even thousands of parasites (including all stages) per animal in moderately infested areas, it is evident that the loss of blood per head can reach between 40 and 50 l per year, a figure which is much increased if a high rate of infestation exists.

Other authors, among them Standfast and Dyce (1968) in Australia, have shown a loss of 166 ml per day of blood in animals with an infestation ranging from medium to high, and this would in turn lead to a loss of 60 l of blood a year.

Regunega (1972) published the results of a trial, in which he compared the weight gain of 10 cows infested with *Boophilus microplus* with an equal number of control animals free of parasites; after approximately 4 months (130 days), these control animals had gained 20 kg, while those infested by *B. microplus* had lost between 4 and 9.5 kg.

In milk production, those animals free of parasites had produced 19 to 42% more than the infested ones.

In summary, estimates in Australia, as well as in Mexico and Argentina, indicate losses caused by *Boophilus microplus* ranging from 40 to 50 kg per head a year. We must add to this figure other additional losses such as the increased susceptibility of parasiticized animals to other diseases (whether parasitical, infectious, or nutritional), the reduced fertility observed on parasitized herds – thus the lowered production of calves – and finally the longer period of time necessary "to get an animal ready" for its slaughter.

Another important factor to be taken into account when evaluating the losses of production is the fact that cattle cannot be genetically improved due to the higher susceptibility towards ticks of the more specialized breeds.

2 Transmission of Diseases to Hematozoons

In the Argentine Republic and in general throughout Latin America, *Anaplasma Marginale, Babesia Argentina* and *Babesia Bigemina*, together with all the sequela of pathological disturbances, cause great losses in farming: not only in the form of variable mortality rates, but in increased expense for their control.

3 Losses Suffered by the Tanning Industry

Not only the damages – of great importance to endemic areas – caused to hides by the bite of the tick itself and by the secondary infections, myasis, etc. must be taken into account, but also those losses caused to the tanning industry as a direct consequence.

Boophilus microplus can be found in an area delineated and between parallels 32°N latitude and 35°S latitude which includes important cattle areas of Central America, South America, Africa and Oceania (Fig. 1.1).

In the latter, specifically in Australia, R. H. Wharton, an acknowledged expert on this subject, defined the common cattle tick as "our most expensive, conspicuous, political and frustrating ectoparasite after it was introduced by way of Java about a century ago" (Wharton 1975).

Fig. 1.1. Geographical distribution of *Boophilus microplus*

As regards the American continent, Lombardo, in his work *Importancia Socio-económica del problema de la garrapata en las Américas* (Lombardo 1976), makes an exhaustive analysis of the problem where, from Mexico to Argentina, there exist approximately 250 million head of cattle, of which about 175 million are located within infested areas, the most severely affected being those with tropical climates. The fact that there exist a high demographic index and obvious nutritional problems within said areas means that the protein loss caused by *B. microplus* is severely distressing to the animal and causes a loss not only of milk production, but also of meat.

With specific reference to Argentina, it will be noticed that approximately 20% of its cattle, about 12 million head of a total of 60 million, suffer attack from *B. microplus*. But, although we are dealing with a partial problem, for it is confined to the northeast and North areas of the country, the losses are of such magnitude that man has been fighting for its erradication for a very long time.

In 1972 a thorough study was made of the losses caused by *B. microplus,* (PRACIVE 1972) and the figures published at the Primera Reunión Argentina de Ciencias Veterinarias (PRACIVE) have been the result of many working meetings of experts on the subject.

Based on these studies, adapting some figures to up-to-date prices and adjusting others because of many changes that have taken place during the last eight years, we arrive at the following data at the end of 1980 (Nuñez 1972, unpublished, Ministerio de Agricultura y Ganadería 1980):

	Argentine $ '000	US $ '000
Losses due to mortality and to the reduction of calving time rate	180,000,000	90,000
Costs of treatments (including acaricides, manual labor)	40,000,000	20,000
Budget of the official campaign	4,000,000	2,000
Total	224,000,000	112,000

Everything so far described and discussed confirms the great economic and social importance of damage caused by the common cattle tick, which in its area of distribution includes many developing countries where the concurrent nutritional problems require urgent solution.

Being able to make, although on a minimal scale, some positive contribution in this struggle against hunger, will justify our efforts in producing this book.

References

Barnett SF (1961) Lucha contra las garrapatas del ganado. Estudios agropecuarios, no 54. FAO, Pome

Lombardo R (1976) Importancia socio-económica del problema de la garrapata en las Américas. Organización Panamericana de la Salud. Publ Cient 316:86 – 97

PRACIVE (1972) Panel de garrapata. Prim Reun Argent Cienc Vet Actas 1:55 – 56

Regunega ME (1972) Eficiencia reproductiva del ganado vacuno. Aspectos económicos. 2nd, Simp Prod Anim, Buenos Aires, Argent

Standfast HA, Dyce AL (1968) Attacks on cattle by mosquitoes and biting mites. Aust Vet J 44:585 – 586

Wharton RH (1975) Vet Sci Intersect Symp 46th ANZAAS Congr, Canberra, Aust

II Taxonomy

Ticks are ectoparasites characterized by hematophagia, no matter to which family, genus, or species they belong, in either warm or cold-blooded animals.

From the taxonomic point of view, the following classification is set forth:

Phylum: Arthropods
Class: Arachnida

For the rest of the classification, we use the same as detailed in the chart (see Fig. 2.1), which has been so kindly sent to us by Dr. Harry Hoogstraal (personal communication).

Within the *Acarina* order, ticks are characterized by having the cephalo-thorax and the abdomen intimately fused. The terminology of the bodily regions are the following:

Prosoma: Buccal apparatus
        4 pairs of legs
Idiosoma:
        posterior region

These two terms of Greek origin mean:

Prosoma: *pre* = before
        soma = body
Idiosoma: *idios* = characteristics of

The *prosoma* together with the neck form the *capitulum*. Its shape varies in accordance with genera and species. It may be either rectangular, hexagonal or triangular (Fig. 2.3) and these characteristics are of great help for their identification. The capitulum is joined to be idiosoma by means of a very thin *articular membrane*.

In females, on their dorsal surface, and embedded into two symmetric small fossae on the prosoma, there appear the porose areas (of Berkese), whose shape varies according to the species (see Fig. 2.3). These two areas, which are lacking in males and larvae, show very thin punctuations which may be the external openings of a series of ducts, which some authors believe may correspond to dermic glands; on the other hand, others are of the opinion that these punctuations

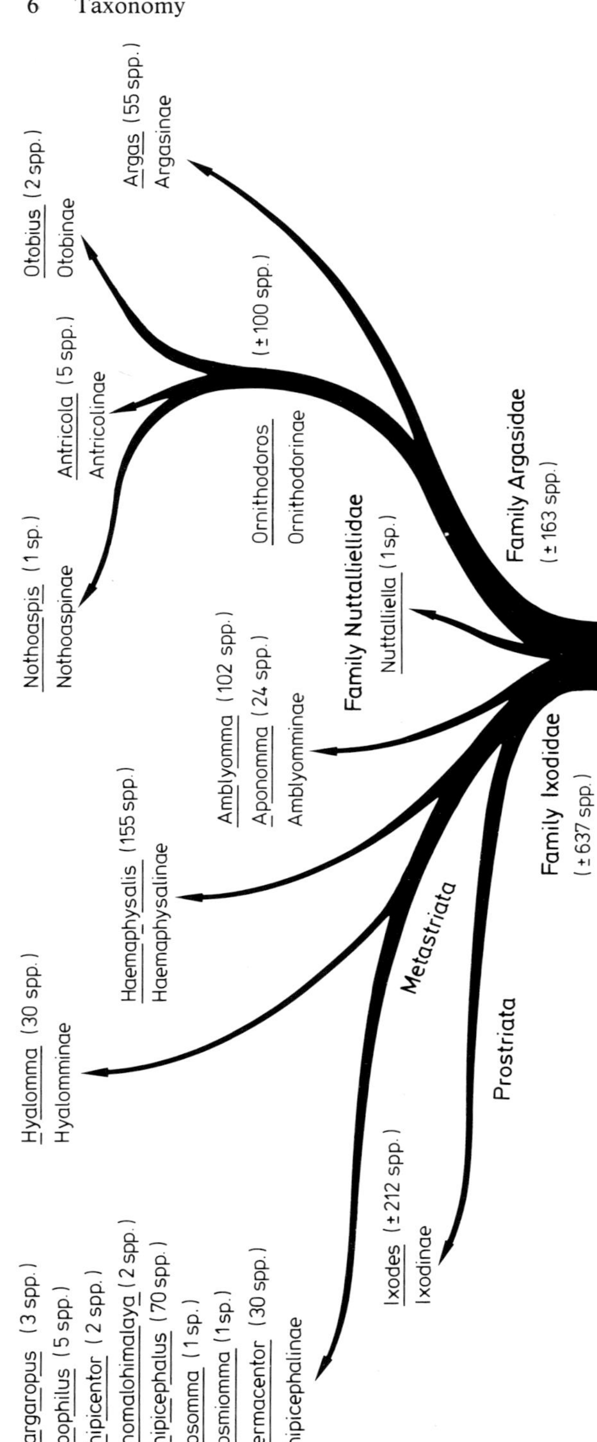

Fig. 2.1. Classification of the Acarina order (Hoogstraal 1981)

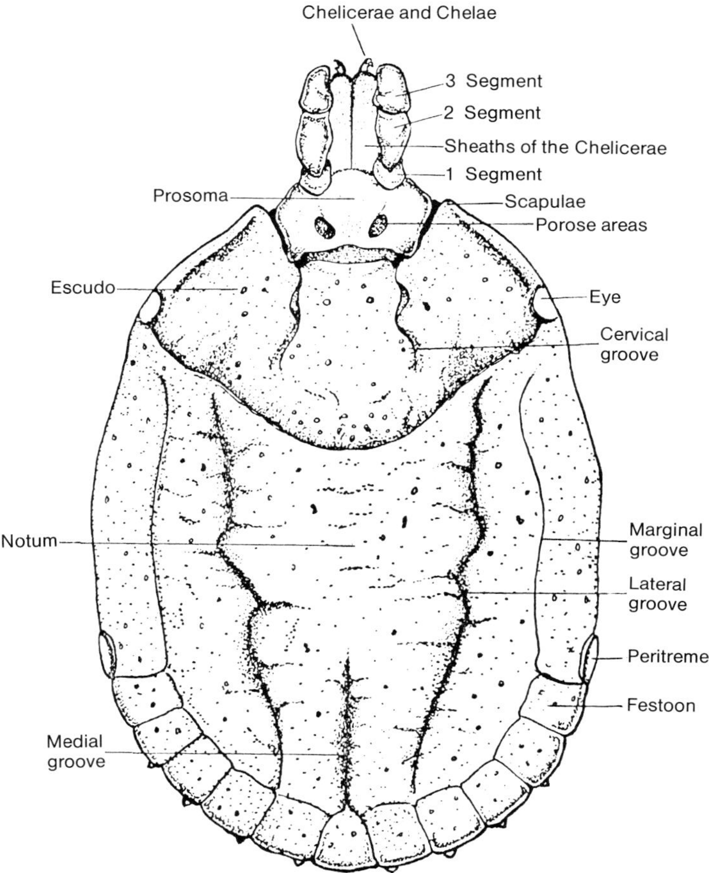

Fig. 2.2. External morphological characteristics of the female *Ixodidae* dorsal view. (Original P. Arrieta)

may be related to the oculoporous nerve, which originates in the dorsal ganglion of the brain.

The *chelicerae,* together with their sheaths, the *hypostome* and the *palps* form the *rostrum.* The former, located over the hypostome, each end in two books called *chelae*; their function is to perforate the skin of the host. In their dorsal area, both are covered by a thin chitinous lamina, thus forming the *sheaths.*

The *hypostome* constitutes the adhering organs, which in turn consist of two symmetric parts, intimately joined between them. The *palps* are located on each side of the *hypostome.* They consist of two tactile organs, whose function is to establish the resistance of the tissue of the host; the palps consist of four segments called *articuli* or segments. The size and shape of these are also used to classify genera and species (see Fig. 2.3).

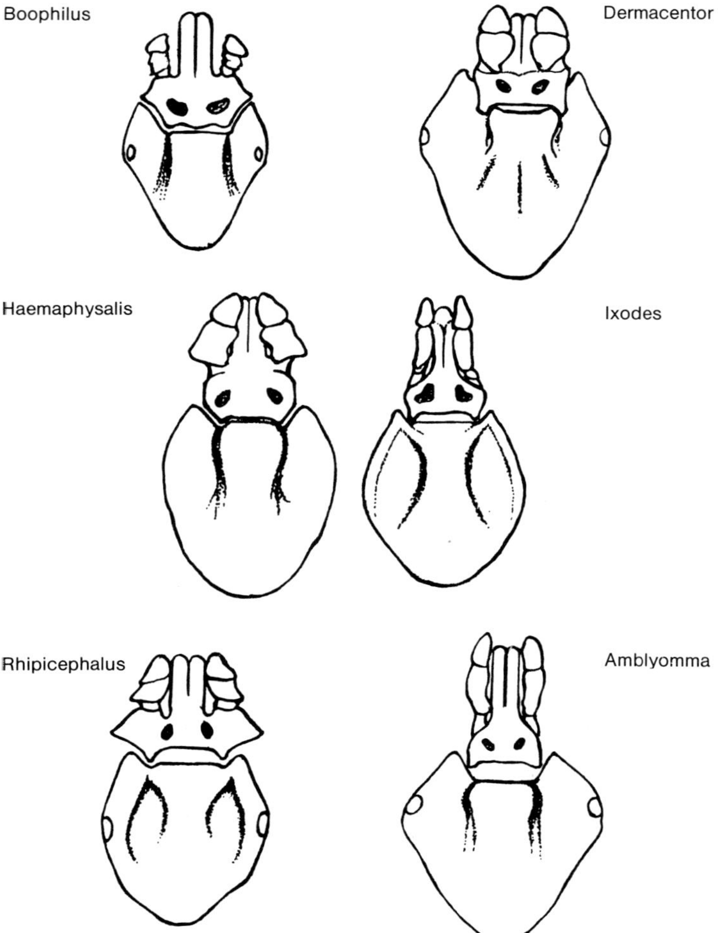

Fig. 2.3. Aspect of the capitulum and shield of females of some *Ixodidae* species (dorsal view). (Original P. Arrieta)

The *basal articulus,* which is joined to the *prosoma,* may or may not have a spur in a ventroposterior position; the *second articulus* may have a similar type of spur, either in the dorsal or lateral position, and this is an important characteristic in classifying genera.

The *third articulus* may be greater or smaller than the second one, and this is also used for classification of the different species.

Finally, *segment IV* is very small. It is located within a deep depression of segment III and its extremity bears a series of erectile hairs (*setae*).

The *shield* is located on the anterior part of the dorsal face of the body. It is formed by chitin, which may be homochromous or ornamented, bearing punctuations of variable sizes and designs of several colours, which together constitute the *ornament*. The shield presents, towards its front, two longitudinal grooves called *cervical grooves,* which originate on the internal angles at the junction with the capitulum. In males, the shield covers practically the whole of the body, leaving only a little marginal zone free; whereas in females the shield is located on the anterior part, keeping its original size.

The anterior border of the shield is notched so as to lodge the basis of the capitulum; this area is known as *emargination.* It has two lateral spurs which form the *scapulae.*

The eyes are located on the borders of the shield, generally on the anterior third. The genera *Amblyomma, Boophilus and Rhipicephalus* have these characteristics, but eyes are lacking in the genera *Haemaphysalis* and *Ixodes* (see Fig. 2.3). That part of the dorsal surface (see Fig. 2.2) uncovered by the shield is called *notum,* which is very much reduced in males, and broad in females. In males the notum is limited by the marginal groove and the lateral borders, but in females it extends from the shield borders to the lateral borders of the body. This notum also presents grooves, the prominence of which depends on the degree of repletion.

In a direction almost parallel to the edges of the body, we can find the *marginal grooves* which, according to the species, may be complete or incomplete, well delineated or limited to a series of punctuations.

Delimited by the marginal grooves and the posterior edge of the body, we find the *festoons,* which have a cellular appearance, and are more or less quadrangular in shape.

On the ventral surface (see Fig. 2.4) towards the front third, and in the area between the *coxae* of the first two pairs of legs, the *genital opening* is located. Its appearance is that of a transverse cleft which may also have a chitinous fold. At this level the *genital grooves* originate and diverge towards the posterior part of the body ending near the festoons, which, in the particular case of *Boophilus* species, are missing.

On the posterior third and on the midline of the body, we can find the *anal opening* or *nephrostome*; it appears as a longitudinal cleft surrounded by two valves which are located within a chitinous ring. The *anal groove* and its extremities, whether convergent or divergent, are also found here, and according to the genera, may be located anteriorally or posteriorly to the *nephrostome*.

The *anal marginal groove* originates at the anal groove and extends to the *central festoon*. Behind the *coxae* of the fourth pair of legs, and in a ventrolateral position on both sides of the body, the *spiracles* open. These correspond to the external openings of the respiratory system. The spiracles are bounded by a *sclerite,* the shape of which can be either round, oval or elongated. This structure is called the *peritreme* or *stigmal plate.*

In males of the genera *Boophilus microplus* and *Hyalomma,* there exist on both sides of the anus the *adanal plates,* which are formed by a thickening of the cuticle; located beside the adanal plates are the accessory ones.

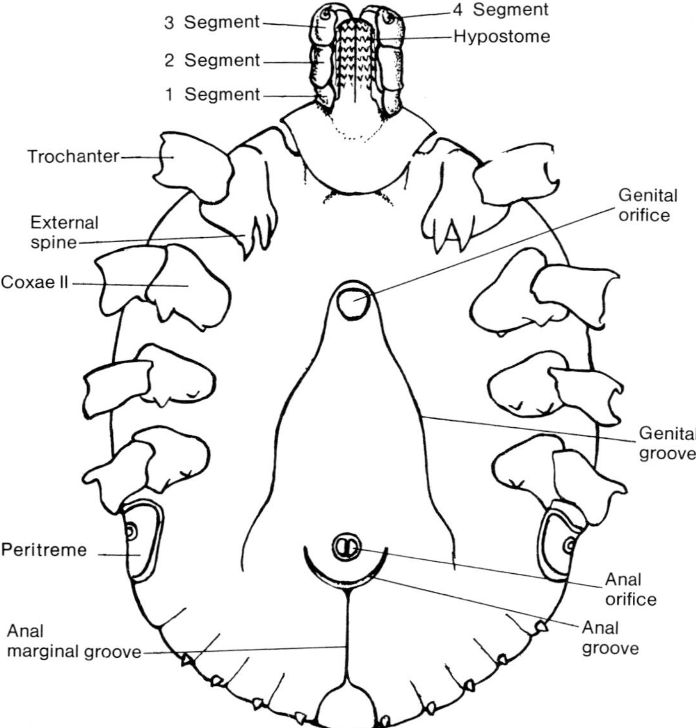

Fig. 2.4. External morphological characteristics of the female *Ixodidae* (ventral view). (Original P. Arrieta)

The legs, of which there are four pairs in the adult tick, are somewhat unequal: the second pair is the shortest one, but the fourth is the strongest and the longest one. They are composed of six articuli, the first one being the *coxa,* the characteristics of which (dentition, tuberosities or spines) are important for the classification of the different genera and species. The second articulus is called the *trochanter,* the third one *femur*; the fourth, *tibia*; the fifth, *pretarsus*; and the sixth, *tarsus*. Located on the tarsus of the first pair of legs we can find Haller's organ, located with a small depression and which, along with the hairs that surround it, form a sense organ. Each tarsus ends in *two nails* plus an intermediate formation acting as an adhesion organ, called the *caruncula.*

External Morphological Characteristics of Argasidae

Ticks with no dorsal shield, also called soft ticks, are included within this family. They lack any kind of pigmentation.

We are going to draw a brief outline about this family with the purpose of describing its most outstanding morphological characteristics, in order to establish the main differences as regards *Ixodidae*.

In this type of tick, less developed than the previous, the prosoma is generally described in the same way as in the case of *Ixodidae,* but differs in that the prosoma has a ventral and subterminal location, and in larvae, it is always terminal.

The most outstanding external morphological characteristics are outlined as follows:

Cuticle: It is striated and mamillated in a regular way. It may present *small fossae,* either oval or circular, called *foveolae,* the number and location of which varies.

Genital Orifice: It is located between the coxae of the first pair of legs, and its shape varies according to sex: in males it is of a semilunar or horseshoe shape, and in females it looks like a transverse depression.

Peritremes: These are located on both sides of the body among the coxae of legs III and IV, being of a circular or semilunar shape.

On a level with the coxae of the first pair of legs, an opening is described which corresponds to a pair of glands which secrete the so called *coxal fluid.*

Pedipalps: Segment IV is equal in size to the other three. Likewise must be noted the sexual dimorphism, which is difficult to perceive, as well as the presence of two or more stages within the life cycle.

Internal Morphological Characteristics of Ixodidae and Argasidae

As exclusive hematophagous parasites, the analysis of the internal disposition of ticks belonging to this suborder is of great importance due, among other reasons, to the fact that certain protozoa, rickettsiae and viruses, fulfill part of their life cycle at this level.

The opportunity is taken here to give a general description of the different systems, so as to analyze them further in detail in the following chapter, where the genus *Boophilus* is dealt with specifically (see Curtice 1981 as Boero (1957) mentioned, or according to Lombardero (1983): Lahille in 1905 changed the genus that Canestrini gave in 1887, so the correct definition is *Boophilus microplus* (Can.).

A Digestive System

The *bucal opening* is located over the hypostome and is limited dorsally by the external sheets of the chelicerae; the excretory ducts of the *salivary glands* open

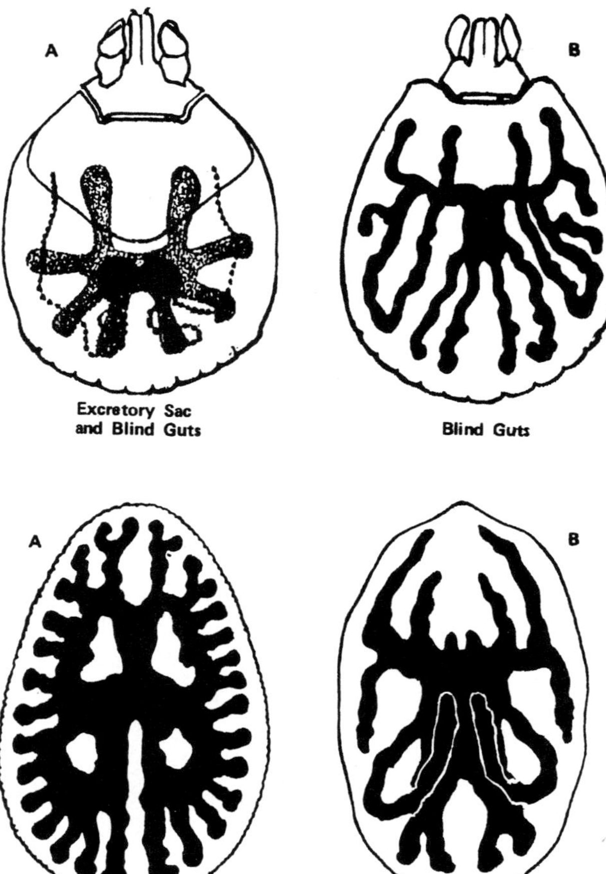

Fig. 2.5A, B. A *Ixodidae* larvae. Excretory sac and blind guts. **B** *Ixodidae* adult. Blind guts (Original P. Arrieta)

Fig. 2.6A, B. A Argas spp. Blind guts. **B** *Ornithodorus* spp. Blind guts (Original P. Arrieta)

into it. A broad and muscled *pharynx* follows, which is transformed into a simple *esophagus*. The esophagus in turn, empties into the *stomach*, which is very dilated; from here two *blind guts* originate. These blind guts have a very characteristic movement, which can be seen in those stages which are not highly chitinized.

B Excretory System

It is composed of *excretory tubules* or *Malpighian tubules,* both of which are characterized by presenting *flexures* (see Fig. 2.6). Their movements are similar to those of the blind guts. Both tubules end in a big dilatation called the *excretory sac*, which can be seen on the ventral side, like a whitish stain (see Figs. 2.5A and

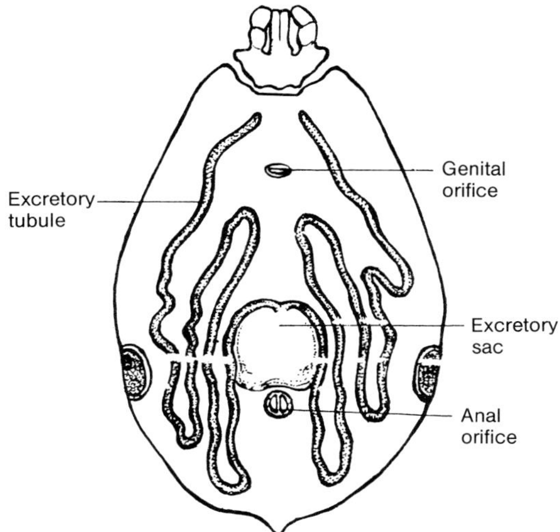

Fig. 2.7. Excretory system of *Ixodidae*

2.7). This system, thus structure, voids its excretory products through an orifice, which has already been described, namely the *anal opening* or *nephrostome.*

C Respiratory System

This starts with small diameter tubules which are included in the coelomate fluid. The diameter of these tubules is regulated by large endothelial cells, thus allowing gaseous exchange.

These tubules, after presenting various branches and growing progressively larger in diameter, and forming two *tracheae*, which in turn open into the internal air sacs which are merely a dilatation of the *spiracles* located on both sides of the body. As has already been described, each *spiracle* is surrounded by *peritremes* or *stigmal plates.*

D Genital Organs

Male genital organs: These are composed of two testes, where the corresponding *vasa deferentia* or *deferent ducts,* also called *spermatis ducts* originate, through which the *spermataphores* are voided. Both ducts join together so that they open finally into the *genital orifice.*
Female genital organs: The oviducts are borne from the extremities of the single *ovary* described. The oviducts in turn open into the genital orifice, which is located between the coxae of the first pair of legs.

References

Arthur DR (1962) Ticks and disease, vol IX. Int Ser Monogr Pure Appl Biol. Pergamon Press, Oxford New York

Boero, JJ (1944) Los ixodídeos de la República Argentina. Boletín técnico de la Dirección General de Ganadería. Minist Agric Dir Sanidad Anim, Buenos Aires

Boero JJ (1957) Las garrapatas de la República Argentina (*Acarina-ixodoidea*). Univ Buenos Aires, Dep Ed

Boero JJ (1967) Parasitosis animales. Tomo III. Ed Univ Buenos Aires

Del Ponte E (1958) Manual de entomología médica y veterinaria argentina. Ed Libr Colegio

Hoogstraal H (1973) Acarina (ticks), chap. 5. In: Gibbs AJ (ed) Viruses and invertebrates. Elsevier, North Holland, Amsterdam New York, pp 90 – 103

Lombardero OJ (1971) Glosario de términos parasitológicos. Ed Univ Buenos Aires

Lombardero OJ (1983) Evolución de los estudios sobre la garrapata del vacuno (*Boophilus microplus*) en la República Argentina en los últimos 100 años. Therios Vol 2, 6:32 – 52, Buenos Aires

Superintendent of Documents. US Government Printing Office (1965) Ticks of veterinary importance. Washington DC 20402 – Stock number 001-000-03461-8/Catalog no. A1.76:485

Wellcome Foundation Ltd (1970) Control de las garrapatas del ganado vacuno. Div Cooper, Wellcome Foundation, London

III Morphology and Physiology

Several authors, such as Minning (1934), Bedford (1934), Theiler (1943), Boero (1944), Cooley (1946), Hoogstraal (1956), Arthur (1960), Gothe (1967), Roberts (1970), Nuñez et al. (1972) and Waladde (1976) among others, have described the morphology of the *Boophilus microplus* tick not only including differential characteristics of each of its life-cycle stages, but taking as reference observations achieved with the aid of the optical microscope.

Nowadays, with the advent of the scanning electronic microscope, a deeper and more detailed morphological study of this species of tick can be achieved, enlarging the knowledge on this subject.

Within this chapter *Boophilus microplus* will be described by sections, together with its apparatus and systems. Simultaneously, reference to the particular characteristics of its different life-cycle stages will be made.

Capitulum

Boophilus microplus as a hematophagous parasite, is considered in the Argentine Republic as the principal vector of hematropic diseases such as piroplasmosis and anaplasmosis. For this reason we begin with the description of the capitulum, which is of great importance as related to the feeding habit of this ectoparasite, either from its internal or external morphological standpoint. In turn, we shall consider in general certain particular characteristic of some of its life-cycle stages.

The capitulum consists, as mentioned above, of the prosoma and the neck. Within the former, the following details must be considered.

Chelicerae

These protrude from the dorsal part of the basis capitulum, and consist of a pair of shafts, each one enclosed within one outer and another inner sheath, which are invaginations of the cuticle produced from the anterior part of the basis capitulum (Allen et al. 1979) (see Fig. 3.1).

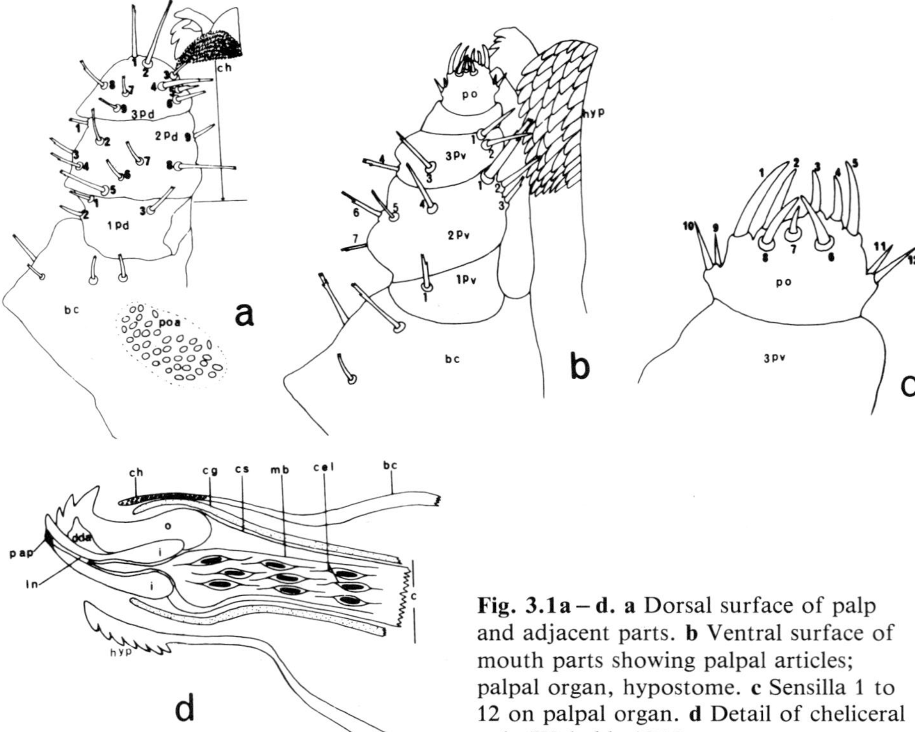

Fig. 3.1a – d. a Dorsal surface of palp and adjacent parts. b Ventral surface of mouth parts showing palpal articles; palpal organ, hypostome. c Sensilla 1 to 12 on palpal organ. d Detail of cheliceral unit (Waladde 1976)

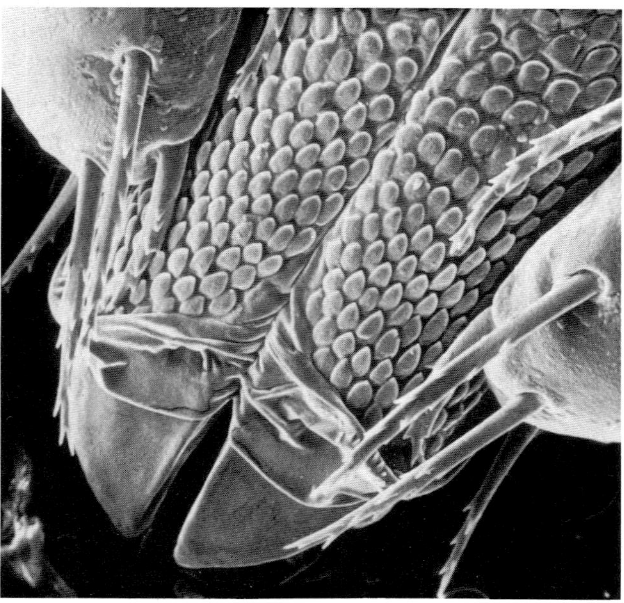

Fig. 3.2. Larva of *Boophilus microplus*, dorsal view. Chelicerae start emerging from their sheaths. 1220 ×

Fig. 3.3. Larva of
Boophilus microplus,
ventral view. Chelicerae
with their chelae and
sheaths. Hypostome with
four rows of teeth and
subterminal dental crown.
1620 ×

Fig. 3.4. Larva of
Boophilus microplus,
anteroventral view.
Chelicerae with chelae
and part of the left palp.
2050 ×

Each shaft bears distally denticulate elements called chelae, which are detailed as shown in Figs. 3.2, 3.3, 3.4.

Each chela is subdivided into one inner and another outer one, each of which in turn bears a pair of teeth; one of these teeth, belonging to the inner chela, shows a papilla located within a shallow depression of a diameter of 2 μm (according to Waladde, 1977). In this same chela, but towards the proximal extremity and in the mid-line, there is an orifice surrounded by a semi-circular edge and which according to S. M. Waladde, would be 9 μm in diameter.

As can be observed in Fig. 3.1 there exists a canal which, originating in the papilla, presents a minor dilatation and then a narrowing. This canal continues within the inner chela, branching at this level into at least three ducts, according to histological studies.

Making use of some of the appropriate coloration techniques, the course of the canal can be observed and, once in the inner part of the chelicera, cells containing well-defined nuclei and surrounded by a membrane can also be identified.

The location of both the papilla and of the orifice in each of the chelicerae suggests that these elements, or at least one of them, act as chemoreceptors. Thus, they have an important role when penetrating the skin of the host, and also when contacting tissues, lymph and blood during the feeding process; and, of course, in the damage that these processes causes.

At the basis of each chelicera a retractile muscle can be observed.

Palps

Palps are another type of appendage which are borne on the anterolateral angles of the capitulum, on both sides of the hypostome. Each palp consists of four semi-cylindrical segments (articuli).

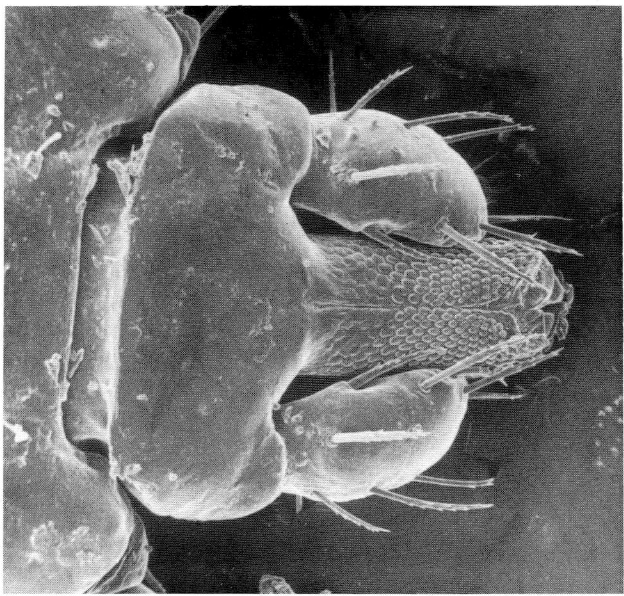

Fig. 3.5. Larva of *Boophilus microplus,* dorsal view of capitulum. 390 ×

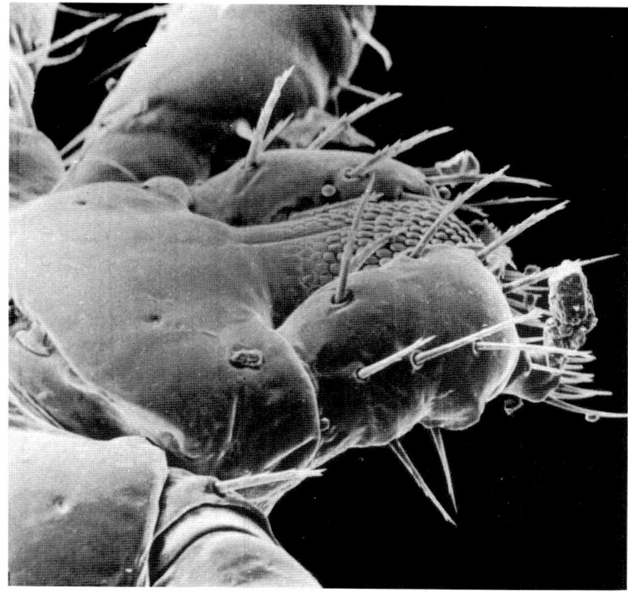

Fig. 3.6. Larva of *Boophilus microplus*, dorsolateral view of capitulum. 490 ×

In larvae, the *palps* are uniform on their dorsal face, and no remarkable separation can be distinguished between segments I and III (see Figs. 3.5, 3.6 and 3.9).

Towards the mid-line on each of the palps, a pair of multiserrated setae can be observed, and, mid-dorsally, another four.

No junction between segments I and II can be observed on the ventral face of the palps; but, in fact, it is possible to notice one, between segments II and III, where the latter thickens. Towards the distal end and middle of segment III, there appears an interarticular membrane which joins this segment to the fourth one.

On the ventral face of palps I – III, three setae can be differentiated, as observed in Figs. 3.7 and 3.8: on the mid-line, a greatly developed and multiserrated one; laterally, another setae, smaller than the previous one, being serrated only on one of its edges; and the third one, which is very small, is located on segment III.

Finally, on the distal end of segment IV of each palp, there appears a group of nine setae, the length of which, according to Waladde (1976, 1977), is 12 – 20 nm. This same author describes a terminal pore within this group of setae, a fact which would indicate that these elements act as chemoreceptors, thus allowing the ectoparasite to analyze the skin of the host before feeding.

If we analyze the palps of the nymph from their dorsal face, segment I lacks any remarkable separation from segment II, as well as any kind of setae.

On the rest of the palps, the setae are present in an equal number and are located in the same way as in the larva, but they are morphologically different in that all of them are straight and have a sharp end.

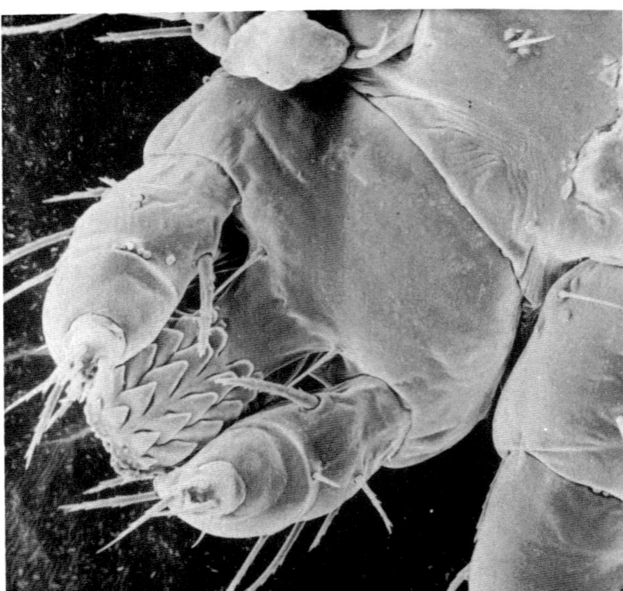

Fig. 3.7. Larva of *Boophilus microplus*, ventral view of capitulum. 490 ×

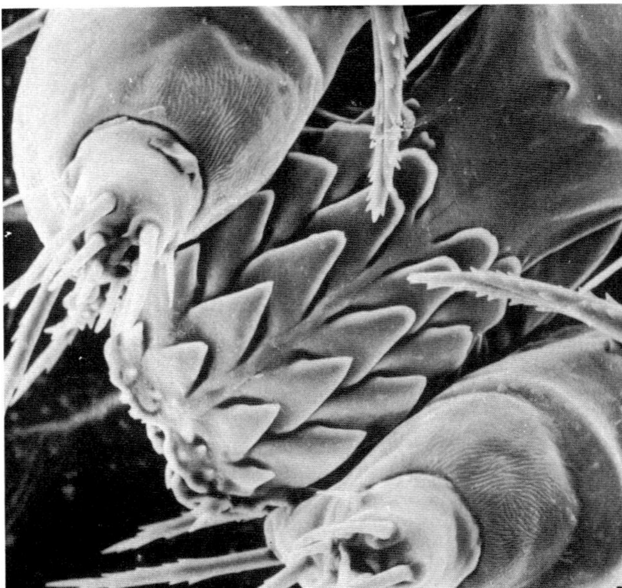

Fig. 3.8. Larva of *Boophilus microplus*, ventral view. Palps. Junction between segments III and IV. At the centre, the hypostome. 1150 ×

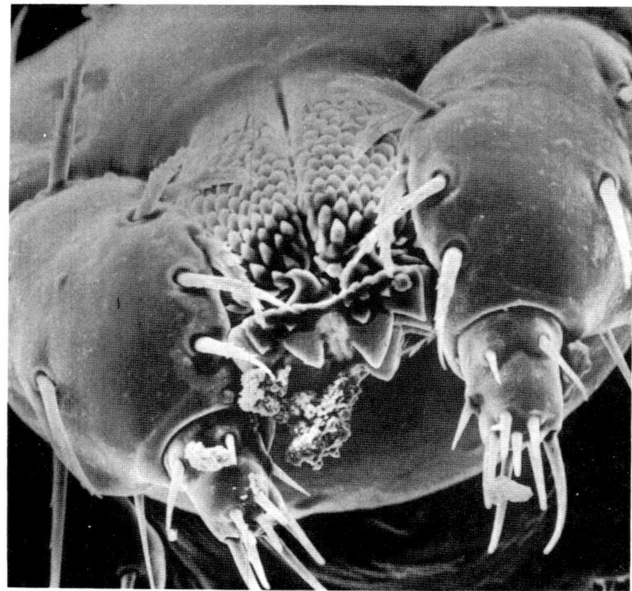

Fig. 3.9. Larva of *Boophilus microplus*, anterior view of capitulum. 820 ×

Ventrally, segment I is separated from segment II by a transverse ridge. Both segments I and III lack setae, whereas segment II bears 3, one of which towards the middle is described as longer, firmer and sharper.

Certain particularities are observed in the adult male in which no clear junction appears between segments I and II dorsally on the palps, but on the base of the latter a well-defined ridge is observed. As can be seen in Fig. 3.10, a posterior ridge bearing a well-defined spur is present ventrally on palps I and III.

Hypostome

This organ for adhering or attachment is located on the ventral portion of the capitulum between the two palps, and consists of two pieces joined on the mid-line. Its size varies from one stage to the next; thus, for example, Londt and Arthur (1975) give the size of the larva from its distal end to the retrohypostomic setae as ranging between 0.097 and 0.11 mm.

In general, the hypostome is dorsoventrally flattened, and its anterior border rounded. It is characterized by presenting in a subapical position a *dentate crown,* and behind it, rows of teeth, the number and shape of which vary in the different life-cycle stages (see Figs. 3.3, 3.7, 3.10 and 3.11).

Thus, in the larva four rows (two in each of its parts) of 5 – 6 teeth each, of which those located towards the basis hypostome are more rounded than the rest.

In the nymph the formula is: 3/3 rows of 5 – 7 teeth each; and these teeth are also sharper distally.

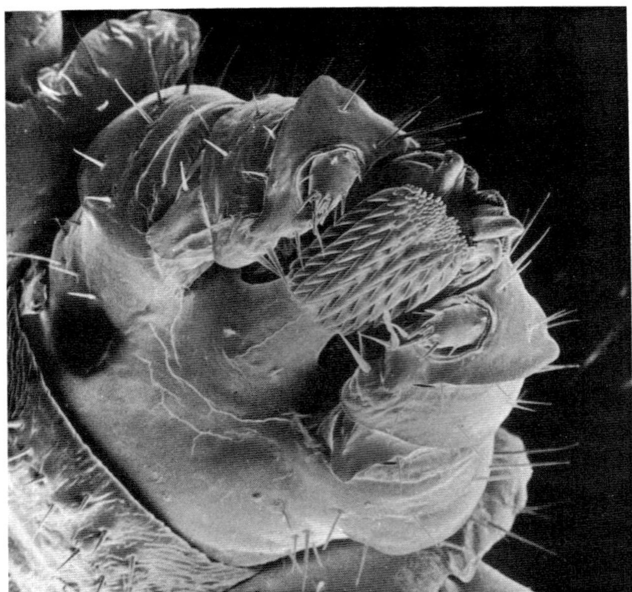

Fig. 3.10. Adult male of *Boophilus microplus*, ventral view of capitulum. 160 ×

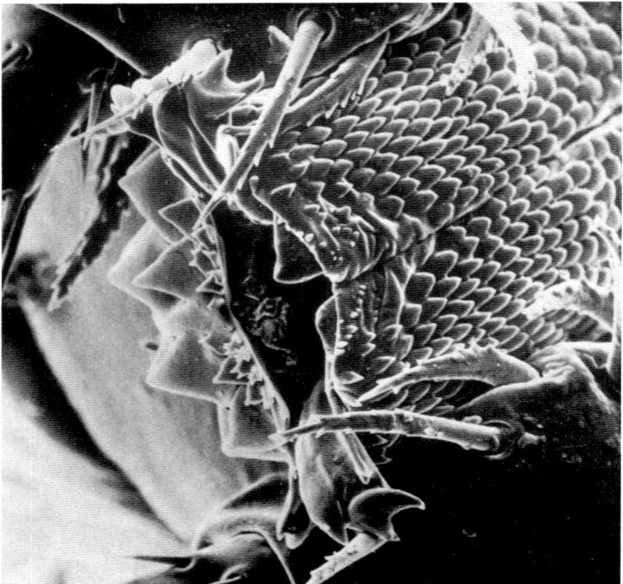

Fig. 3.11. Larva of *Boophilus microplus*, anterior view. Distal end of capitulum. 1180 ×

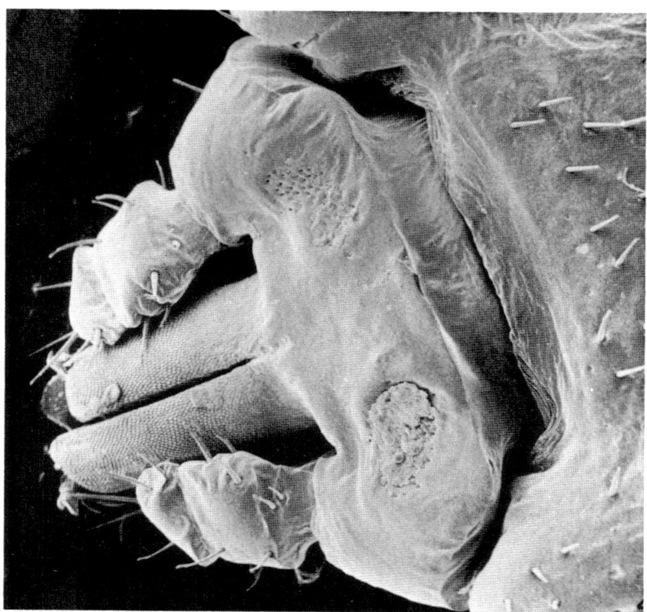

Fig. 3.12. Semi-engorged female of *Boophilus microplus*, dorsal view capitulum. 140×

Both in undistended and semi-engorged females, the formula is 4/4 rows of teeth with 11 – 12 elements each.

In adult males, the dentate crown occupies a considerable portion (approximately 1/6 of the dentate surface of the hypostome); here 4/4 rows of teeth are also present with the following characteristic: the last two rows, on both sides and towards the lateral aspect, consist of 8 – 9 teeth, whereas the other two consist of only 5 – 7 teeth.

Table 3.1. Capitulum of *Boophilus microplus* – some measurements

Stage	Length (from anterior end of palp to posterior edge of neck, mm)	Width (on its basis, mm)	Reference
Larva	0.15	0.18	Cooley (1946)
	0.13	0.17	Roberts (1969)
Nymph	0.24	0.30	Cooley (1946)
	0.22 – 0.26	0.28 – 0.30	Arthur (1962)
	0.25 – 0.28 or 0.30		Londt and Arthur (1975)
Adult male	0.33 – 0.40	0.40 – 0.49	Cooley (1946)
	0.30 – 0.37	0.40 – 0.50 (on its central part)	Arthur (1962)
Undistended female	Approx. 0.45	0.62 – 0.66	Cooley (1946)
	0.40 – 0.56	0.56 – 0.70	Arthur (1962)

Considering the capitulum as a whole, it can be appreciated that its shape varies from one stage to the next. But, in general terms, it can be said that dorsally its lateral borders are rounded, whereas the posterior one is straight. Now, ventrally, the posterior border is rounded and the capitulum as a whole presents a slight concavity; always on the ventral surface, and behind the hypostome on both sides of the mid-line, there exists a pair of setae accompanied by two depressions (retrohypostomic setae).

The *porose areas,* which have been described already in the previous chapter, appear again at this level among certain other details which can be observed in Fig. 3.2.

To round off the description of the capitulum, some measurements according to data given by different authors are given in Table 3.1.

Legs

The first segment that constitutes them corresponds to the *coxae* which are only visible ventrally.

In the larva, the coxa of leg I bears a posterior well-defined spur, as can be observed in Figs. 3.13, 3.14 and 3.15.

This spur appears in a more reduced size on the coxa of leg II, and on leg III it appears like a small mamilla. The number of setae present at this level is 3 on leg I and 2 on the rest. In the adult stages, it can be observed, as in the particular case of the semi-engorged females (see Fig. 3.16), that the shape and size of the spur changes from leg I to IV. Likewise, in adult males, certain particulars can be noticed, as shown in Fig. 3.17; on the coxa of leg I appears craneally a greatly developed spur, which is why it can be seen dorsally; caudally, there appear two greatly developed appendages or spines, separated by a well-defined V-shaped cleft.

These latter are less observable as we proceed to coxae on leg IV. The rest of the segments, from mid-line to lateral, correspond to trochanter, femur, tibia, protarsus and tarsus; the latter ends in two claws and pulvillus or caruncle. Their characteristics can be observed in Figs. 3.18, 3.19, 3.20 and 3.21 A – B.

The *tarsus* is an element which deserves a detailed description, together with Haller's organ, which is located on it.

Waladde (1976) made a complete report on this subject; his studies are taken as reference. This author divides the tarsus into three zones: distal, medial and proximal (Fig. 3.22).

The first of these comprises the protarsum (*pt*), the pair of claws (*cl*) and the pulvillus (*pul*). Following and projecting onto the protarsus, there appear 3 pairs of setae identified as *dd, dl* and *dv.*

The medial zone includes a series of setae subdivided into three groups, namely: 13 medial dorsal (*md*), 8 medial lateral (*ml*) and 10 medial ventral (*mv*).

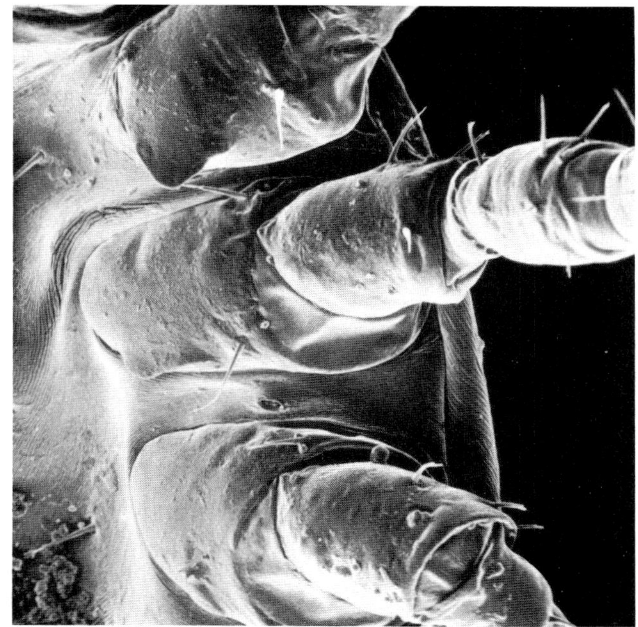

Fig. 3.13. Larva of *Boophilus microplus,* ventral view. Coxa from legs I – III, left side. 400 ×

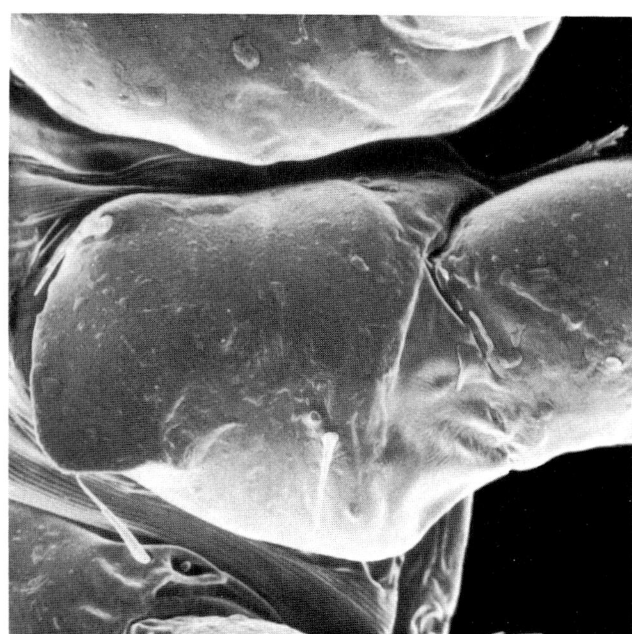

Fig. 3.14. Larva of *Boophilus microplus,* ventral view. Coxa of leg I, left side. 800 ×

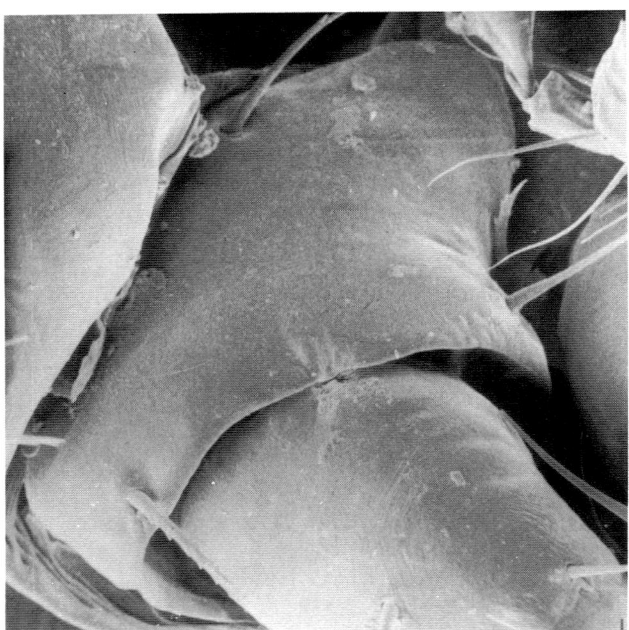

Fig. 3.15. Larva of *Boophilus microplus*, ventral view. Coxa of leg I, right side

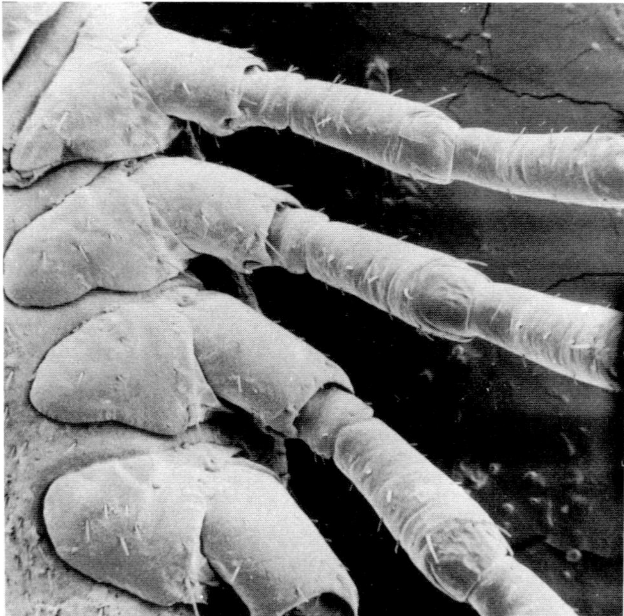

Fig. 3.16. Semi-engorged female of *Boophilus microplus*, ventral view. Coxae I – IV, left side, 70 ×

Fig. 3.17. Adult male of *Boophilus microplus*, ventral view. 40 ×

Fig. 3.18. Semi-engorged female of *Boophilus microplus*, ventral view. Complete trochanter and part of the femur, left side. 250 ×

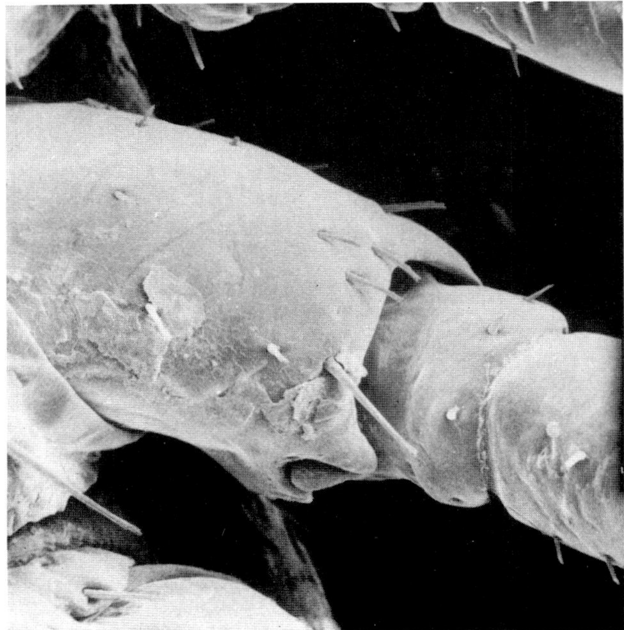

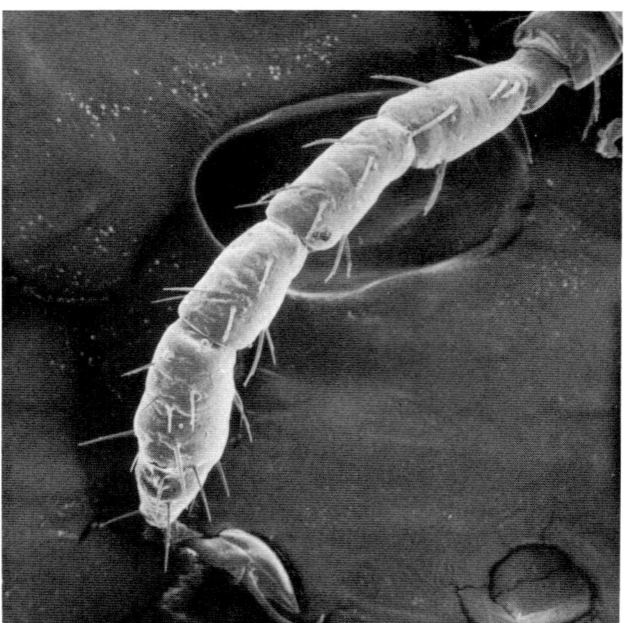

Fig. 3.19. Larva of *Boophilus microplus*, dorsal view. Detail of leg I, right side (trochanter − Pulvillus). 240×

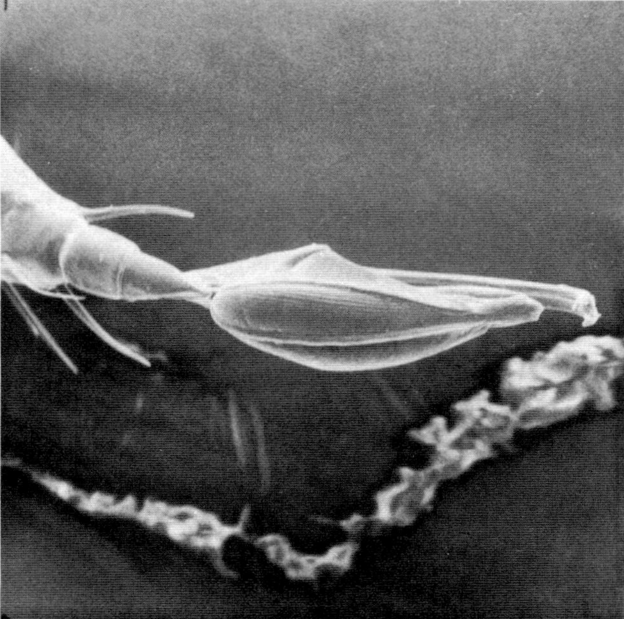

Fig. 3.20. Larva of *Boophilus microplus*, detail of pulvillus. 640×

Fig. 3.21 A, B. Semi-engorged female. Detail of claws. 820 ×

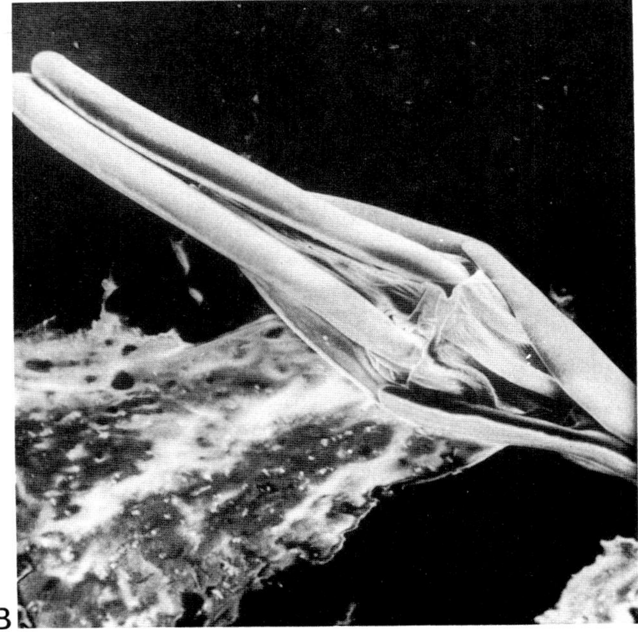

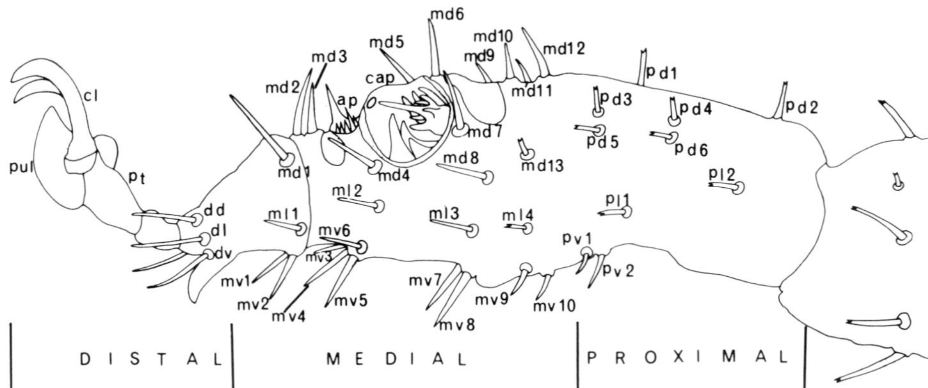

Fig. 3.22. Tarsus of an anterior leg of *Boophilus microplus* (Waladde 1976)

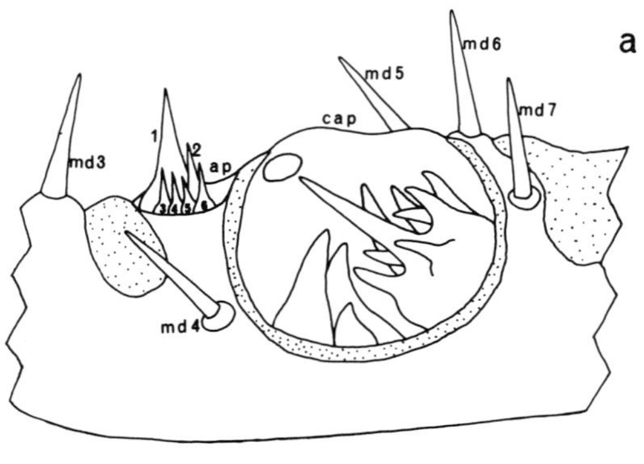

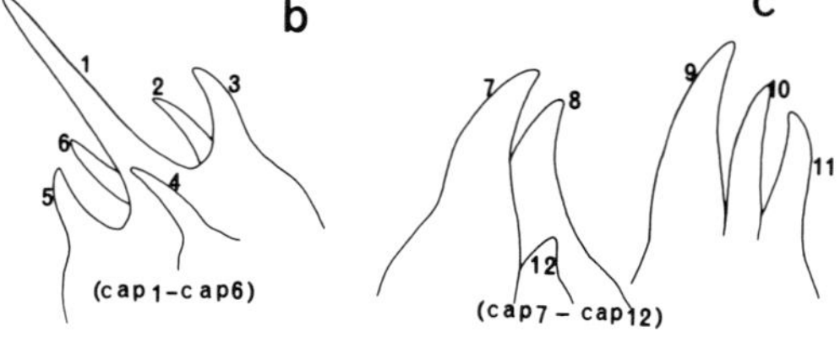

Fig. 3.23a – c. a Anterior depression (*ap*) bearing setae 1 – 6 and posterior capsule (*cap*) of Haller's organ; **b** Setae 1 – 6 (*cap 1 – cap 6*) of posterior capsule; **c** Setae 7 – 12 (*cap 7 – cap 12*) of posterior capsule (Waladde 1976)

Likewise, on this level we can find Haller's organ (Haller 1881) which consists of an anterior depression (*ap*) and of a posterior capsule (*cap*) (see Fig. 3.23).

In general, at this level, the setae are characterized by lacking an inserting orifice, and appear as if mounted on a platform. In the anterior depression, there appears a group of 6 setae, among which *ap 1*, of 30 – 36 μm long, protrudes; the rest being from 5 to 3 m long. Taking Waladde's published reports (1977) as reference, the surface of the setae of *ap* would be rugose and full of pores; these two characteristics being typical of chemo- and olfactory receptors (Slifer 1970).

These setae are innervated by many neurons, forming with their long dentrites a large group on the basis of the anterior depression.

The posterior capsule shows an exterior opening of irregular shape, a drawing of which is given in Figs. 3.24, 3.25 and 3.26.

Waladde describes (1976) in the inner part of this capsule, a total number of 12 setae, divided into 2 groups (see Fig. 3.23 A, B and C). The first group consists of 6 setae (*cap* 1 to *cap* 6) which have a common basis; *cap* 1 is the longest, 32 μm in length, its distal end reaching the external opening of the capsule.

The other group consists of another 6 curved setae (*cap* 7 to *cap* 12), the length of which ranges between 27 and 30 μm. Although some of the capsular setae may not be innervated, the vast majority of them have multiple innervation, a fact which has been corroborated by Waladde (1976) by observing with the optical microscope that several cytoplasms of nerve cells and dendrites were located on the basis of these setae.

With the scanning electronic microscope, the setae appear at the external opening of the posterior capsule (see Fig. 3.26) and bear on their surface a series of pores, which confer to this structure a sponge-like appearance. This last finding confirms the idea that some of these setae act as olfactory organs.

In relation to this last point, Binnington and Tatchell (1973) mention that some axons, which supposedly originate in Haller's organ (Tsvileneva 1964) and then, going through the nerves of leg I, run into the olfactory glomeruli of the brain.

The *tarsal gland* is located on the central portion of the tarsus. This gland is described by Chow et al. (1972) with respect to *Rhipicephalus sanguineus* and by Waladde (1976) as regards *Boophilus microplus*. The function of this gland, in accordance with those authors (Chow et al. 1972) is not completely understood, but it is speculated that the gland may be related to the presence of lipidic material located in the empodium, that is (Keirans et al. 1976) in the whole structure forming the end of the acarus legs, because the main secretion of this gland is represented by lipids.

Furthermore, as the flexor tendon is located within the tarsal gland, the latter would produce a lubricant substance, which in turn would allow the tendon to work better.

Finally, another possibility may be that the gland produces sexual pheromones; this may be so because of the similarity of its structure to that of glands that have this specific function, and which have been described by several authors as regards other species.

In general, the tarsus bears 66 setae located in dorsal, lateral and ventral positions within the 3 zones already described. These setae have sharp ends and are orientated towards the distal end of the limb (see Fig. 3.27).

Apparently, and as described by Waladde (1977), these setae act as mechanoreceptors, lacking both terminal pores typical of chemo-receptors and any pores on the rest of the surface, which are characteristic of olfactory receptors.

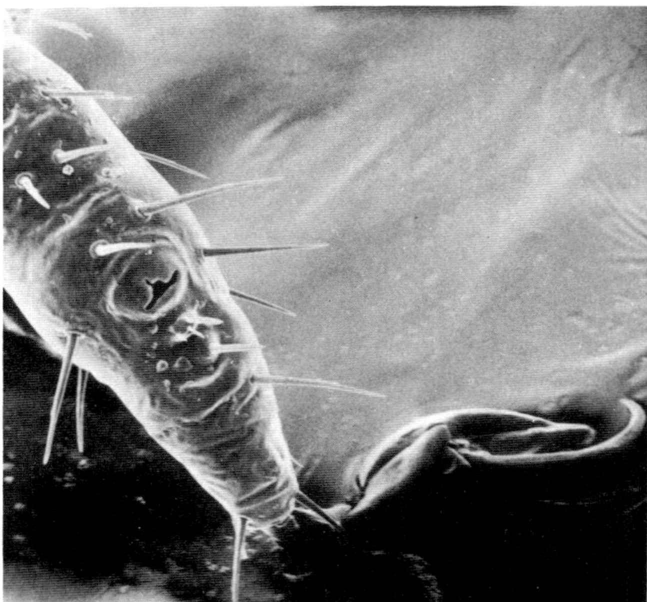

Fig. 3.24. Larva of *Boophilus microplus*, dorsal view. Detail of tarsus, anterior leg. 480 ×

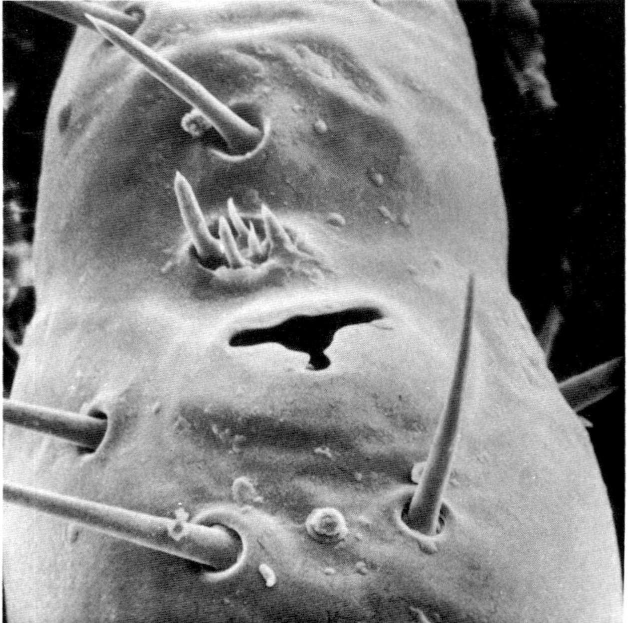

Fig. 3.25. Larva of *Boophilus microplus*, Haller's organ in detail. The anterior depression and the external opening of posterior capsule can be seen

Fig. 3.26. Larva of *Boophilus microplus*. External opening of posterior capsule, 3 setae can be observed therein. 2480 ×

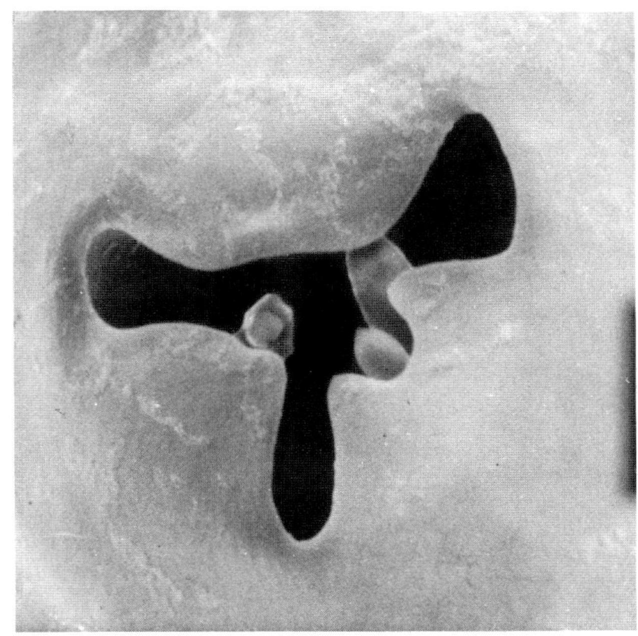

Fig. 3.27. Larva of *Boophilus microplus*, lateral view. Detail of tarsus of leg I. 820 ×

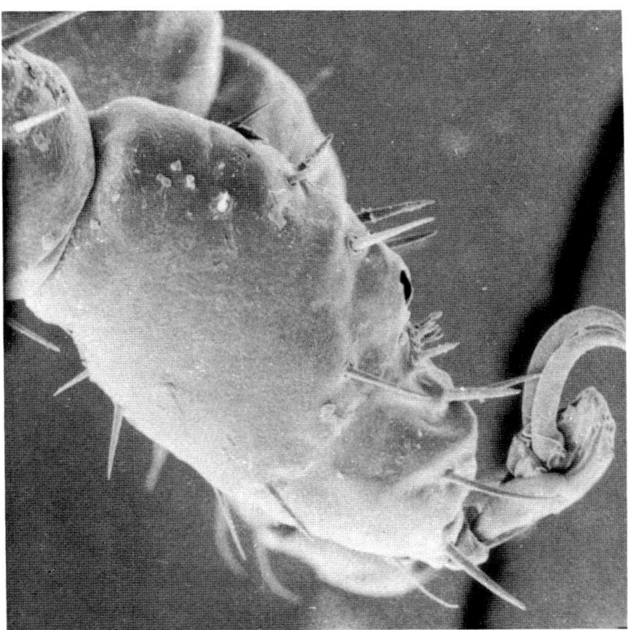

Dorsal Shield

As mentioned, the *shield* is dorsally located on the anterior part of the body. It is, in general, shiny, plain and dark brown.

In larvae, the shield is wider (0.39 mm) than it is long (0.27 mm) (Roberts 1969), and shallow cervical grooves extend in a parallel manner, to the middle of the shield. There appear 3 pairs of setae: Pair I on the anterior part of the shield and towards the lateral of the cervical grooves; Pair II near the lateral border of the shield and on its central portion, and Pair III at the centre and slightly beneath the level of the eyes.

According to Roberts (1969) its length ranges from 0.010 to 0.017 mm. In the nymph, Londt and Arthur (1975), for this structure in unfed ticks, give the following sizes: length taken from the end of scapulae to the posterior border, approximately 0.56 mm; width at eye level, approximately 0.52 mm; length – width relation 1.08 : 1.00. At this stage the border is well delimited, the cervical grooves being shallow and convergent on the first portion. The grooves diverge when they reach the posterior border in a posterolateral position.

The setae are the same as those that have been described for the larva, but Londt and Arthur (1975) describe further a setal pair at the centre of the eyes; another on the anterior portion of the shield between the cervical grooves, and one or two pairs on the scapulae.

A panoramic view of the dorsal shield of the semi-engorged female can be observed in Fig. 3.28.

The sizes of the shield, according to Gothe (1967) for ticks collected in South Africa, would be 1.18 – 1.24 mm long and 0.90 – 1.06 mm wide.

At this stage, the cervical grooves, although superficial, appear well delineated, reaching the posterior border, as observed in Fig. 3.29, where, in addition, the right eye can also be seen. Here these grooves also tend to converge centrally, and then to run divergently to the posterior border. The setae, situated on scapular level, are short; the length of the rest being quite reduced. The appearance of one of them can be clearly seen in Fig. 3.30.

In adult males, the shield covers the dorsal surface and is rigid. Its appearance can be observed in Fig. 3.31.

The cervical grooves run divergently in an anteroposterior direction, on the level of leg II where the eyes are located, although only rarely apparent.

Posteriorly, there appear two lateral grooves and a central one, which would originate in a circular depression on the level of leg IV and which are described in Fig. 3.32.

The great number of setae, existing all over the dorsum on the level of the elevated zones, can be observed both in Figs. 3.31 and 3.32.

Among the elements composing the ventral surface are included the coxae, which have already been described.

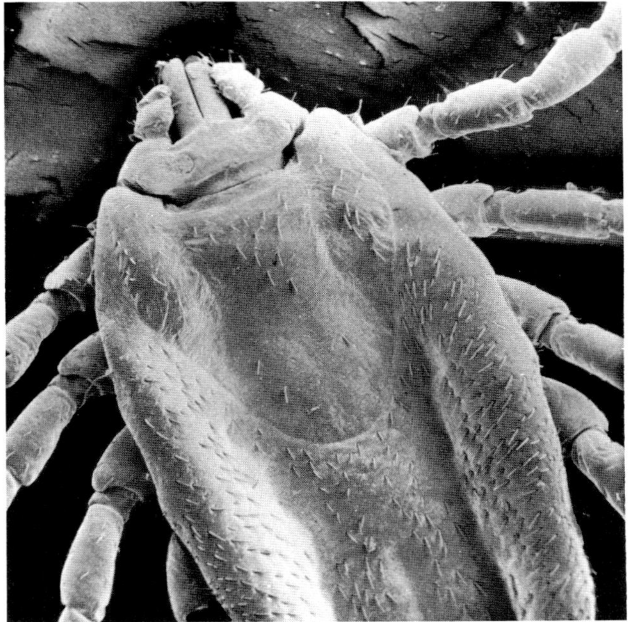

Fig. 3.28. Semi-engorged female of *Boophilus microplus*, dorsal view. 40 ×

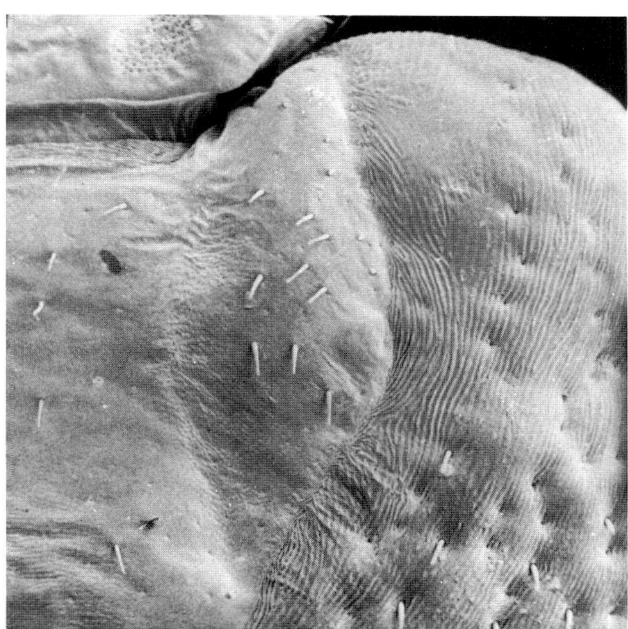

Fig. 3.29. Dorsal view. Detail of half the dorsal shield, right side. 100 ×

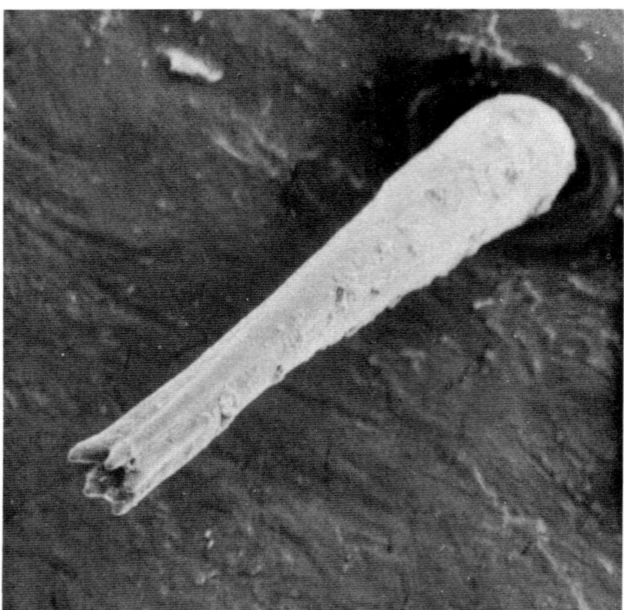

Fig. 3.30. Seta of dorsal shield, near the eye. 2480 ×

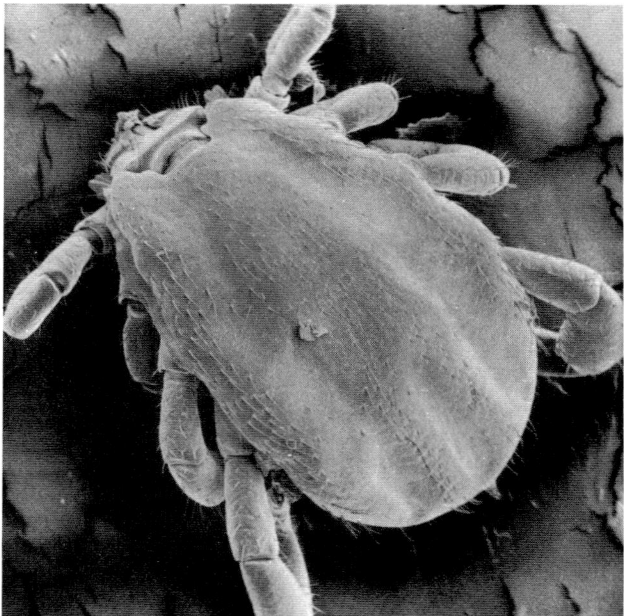

Fig. 3.31. Adult male of *Boophilus microplus*, dorsal view. 40 ×

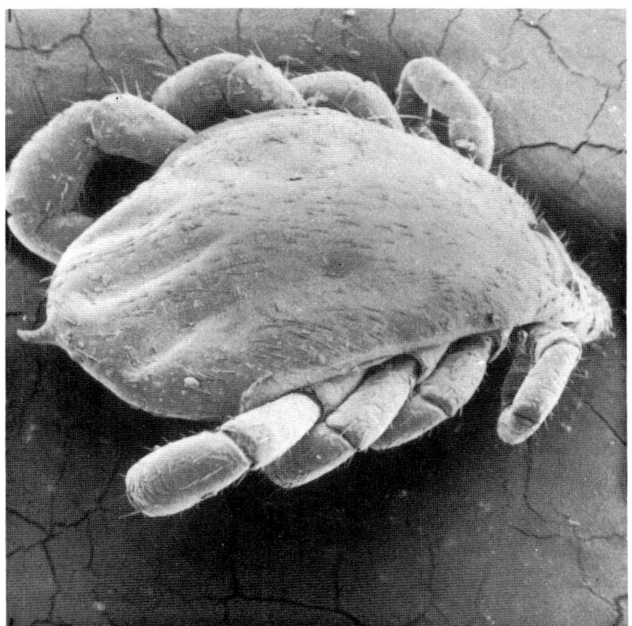

Fig. 3.32. Adult male of *Boophilus microplus*, right dorsolateral view. 40 ×

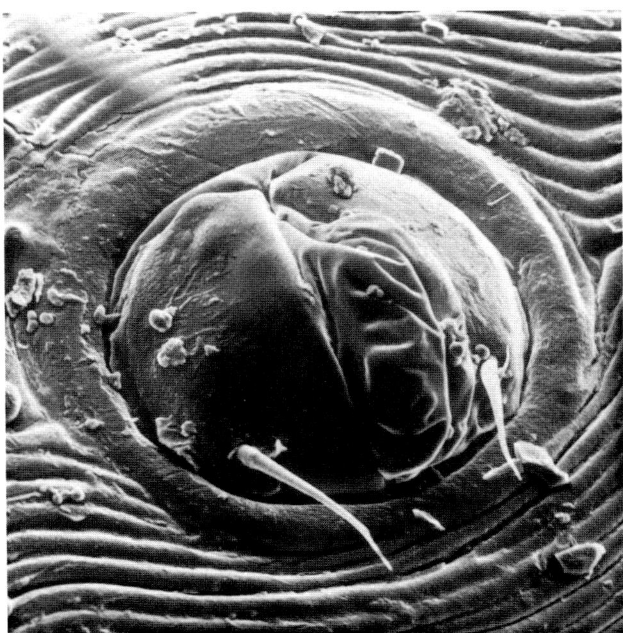

Fig. 3.33. Larva of *Boophilus microplus*. Anal opening. 1600 ×

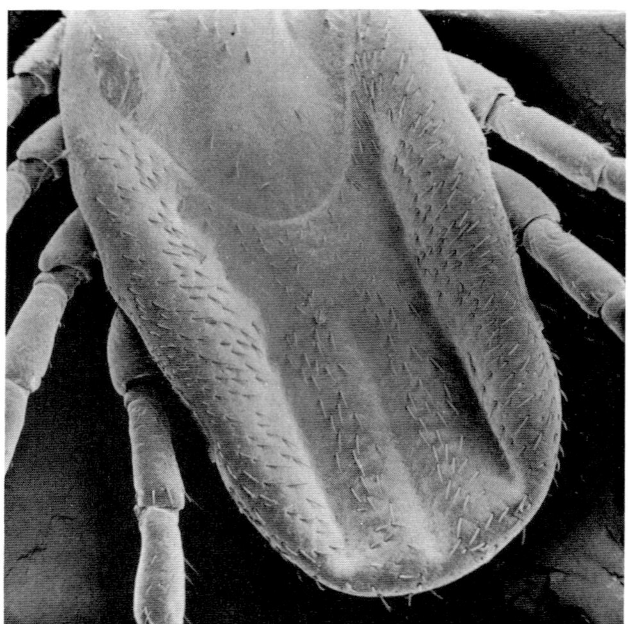

Fig. 3.34. Semi-engorged female of *Boophilus microplus*, dorsal view. 40 ×

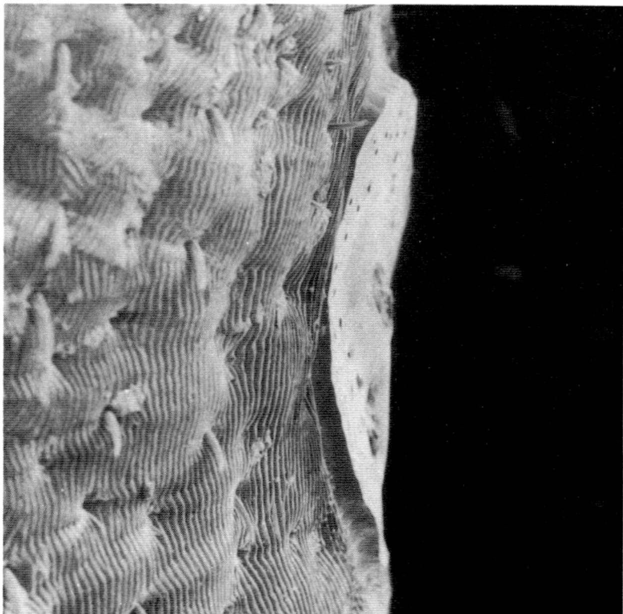

Fig. 3.35. Semi-engorged female of *Boophilus microplus*, peritreme. 240 ×

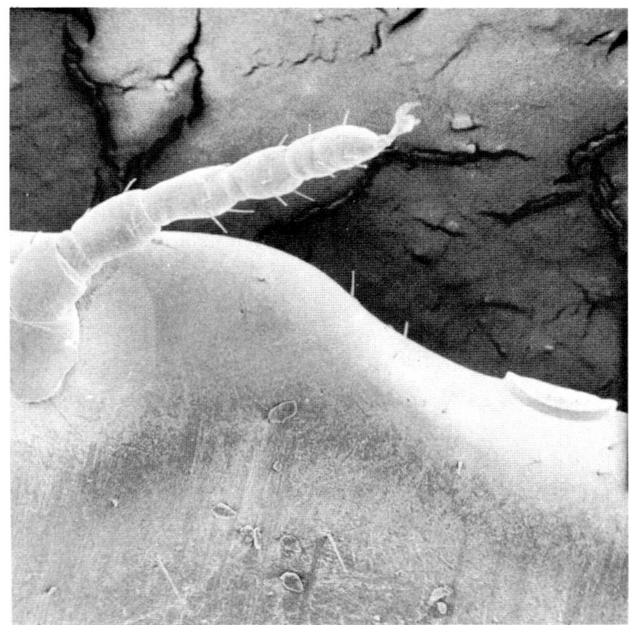

Fig. 3.36. *Boophilus microplus*, engorged nymphal exuviae, ventral view, right side. Detail of the last leg and of the peritreme. 110 ×

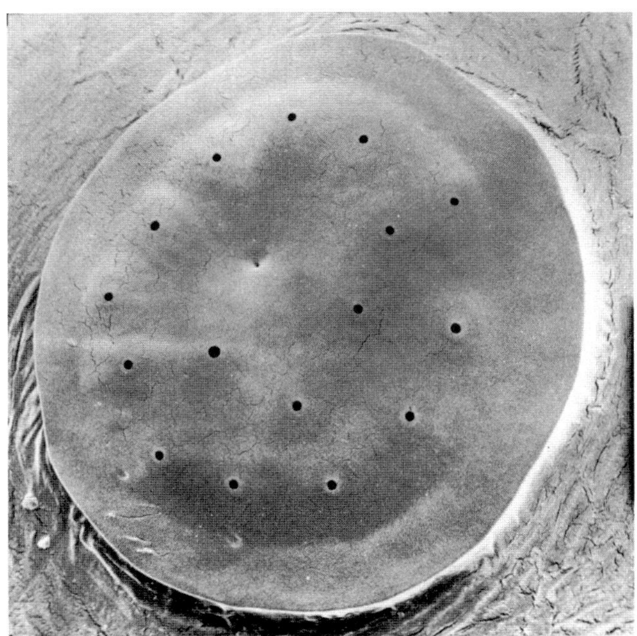

Fig. 3.37. *Boophilus microplus*, engorged nymphal exuviae. Peritreme. 570 ×

Particularly, we note some elements within this surface in the specific case of males, details of which can be observed in Fig. 3.17.

The *genital opening*, as in females, is located between the coxae of leg II; whereas the *anal opening* is located on the posterior third and is surrounded by a chitinous ring bearing, caudally and on the posterior part, two setae (see Fig. 3.33).

The *adanal plates* reach the posterior border, being generally concave at this point, with two spurs or spines; the one located on the mid-line is the longest.

The *accessory plates,* which are generally shorter than the foregoing ones, are located on both sides and laterally to the adanal plates.

The *peritreme* is similar in both males and females. It protrudes slightly from the surface of the body (see Figs. 3.34, 3.35, 3.36 and 3.37).

Digestive System

On the superior surface of the hypostome there exists a furrow covered by a membrane which, together with the external sheaths of chelicerae, dorsally from the buccal opening. Immediately caudally, the labral lobule or labrum is located. This separates the pharyngeal cavity from the salivarium, the latter being the point through which the saliva runs; this, in turn, is secreted by the salivary glands and the labrum serves as a reservoir. Diagrammatically, this may be understood by observing Fig. 3.38.

In unfed ticks, both salivary glands extend to both sides of the body from the scapulae of the shield on the anterior portion and caudally to the level of the peritreme, that is, behind the coxae of leg IV.

These glands have a raceme-like appearance − as can be appreciated in Fig. 3.39 A and B − the main ducts of which empty into the salivarium.

According to Megaw and Beadle (1979), each salivary gland is composed of approximately 400 acini. This author classifies them into three different types.

Those belonging to *type I*, also called non-granular, are located on the anterior portion, and constitute less than 5% of the whole. These acini contain 4 pyramidal cells, separated from the lumen of the acinus by means of a central cell. Acini *type II* constitute approximately 40% of the gland, and are located on the anterocentral portion of the same. These acini are constituted by three varieties of secretory cell, whereas acini *type III* are constituted by two types.

Rovere et al. (1980), from observations with the optical microscope, describe a type of monolayer prismatic cell bearing central or paracentral nuclei without being in state of mitosis, and of cytoplasm optically like that of the type of serous secretion with vacuolations of different diameters and numbers. On the whole, the inner border of each of these cells directly forms the wall of the central duct, as can be observed in Fig. 3.40.

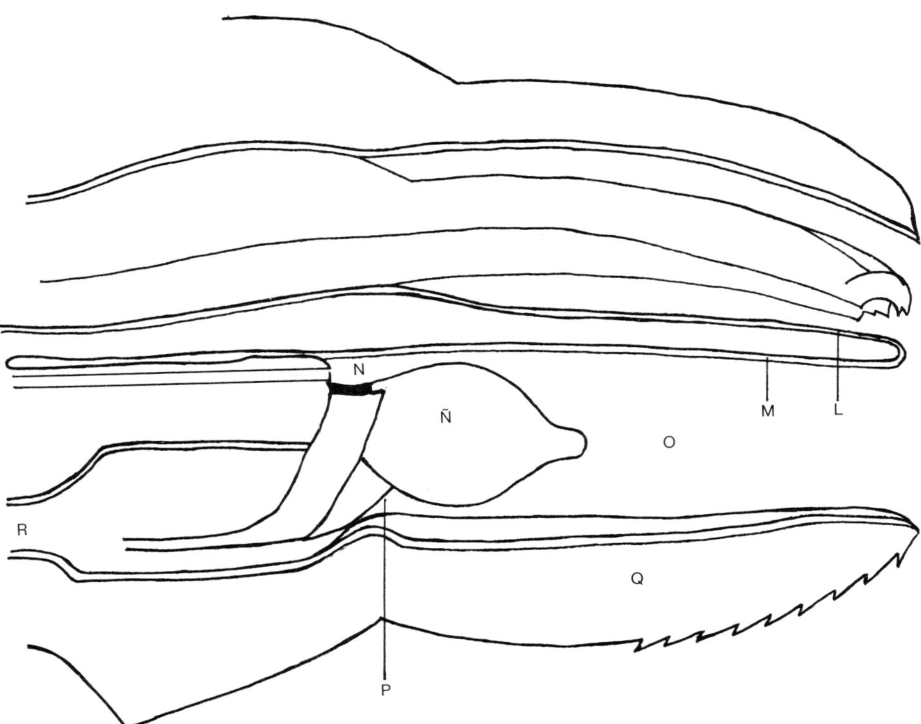

Fig. 3.38. *L* inner sheath of chelicerae; *M* outer sheath of chelicerae; *N* salivarium; *Ñ* labral lobule (labrum); *O* buccal opening; *P* pharyngeal orifice; *Q* hypostome

In these same cells, observed on the scanning electron microscope, the cytoplasmatic membrane harmonically alternated, which run through it (see Figs. 3.41 and 3.42, obtained by SEM).

Likewise, in other zones of these glands, there appear cells very similar to those mentioned above, but more voluminous, as can be seen in Fig. 3.43.

Rovere et al. describe a third type of cell like large gigantocytes (see Fig. 3.44), which also keep the nucleus – cytoplasm relation. Within the latter the presence of a dense granular endoplasmic reticule stands out, a fact which would indicate an active secretory function. Several cells in process of mitosis were also observed among these.

Considering some of the functions which these glands might perform, it is suggested that they may be involved in the production of a hygroscopic fluid which, when making contact with the buccal apparatus, would allow the parasite to take up humidity from the environment, thus facilitating the feeding process (Rudolph and Knulle 1974), in which the fluid would also perform an excretory function voiding the excess of liquids. Likewise, these glands may be related to the regulation of ions, mainly sodium and chlorine from the hemolymph. It has

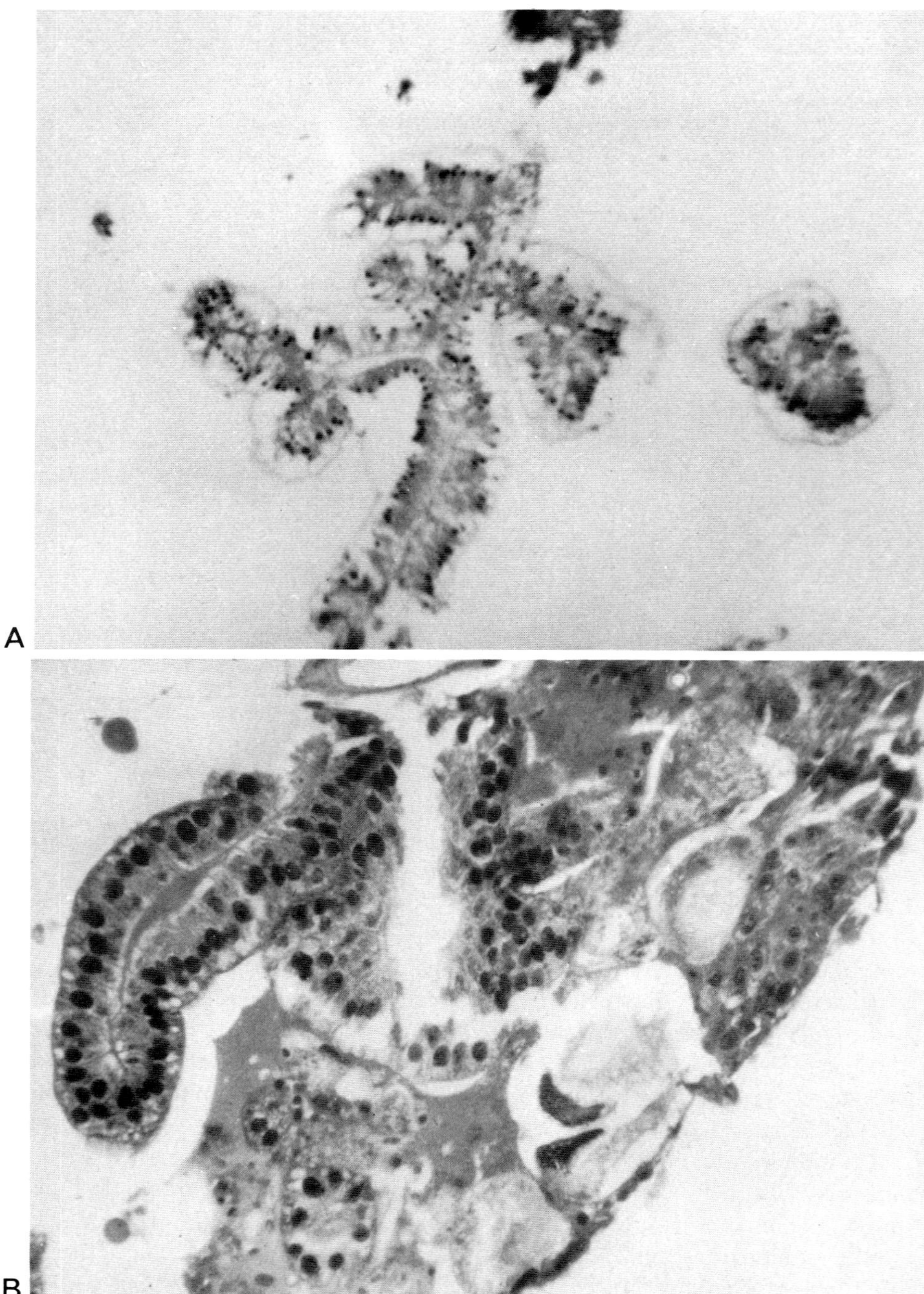

Fig. 3.39 A, B. General view of one of the salivary glands of an undistended female observed on an optical microscope. (Rovere et al. 1980)

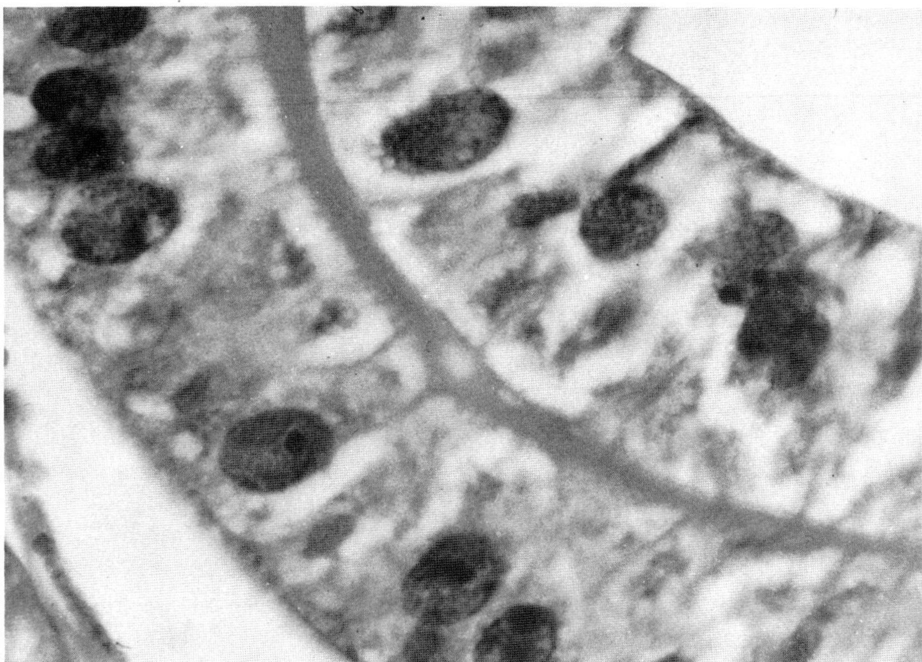

Fig. 3.40

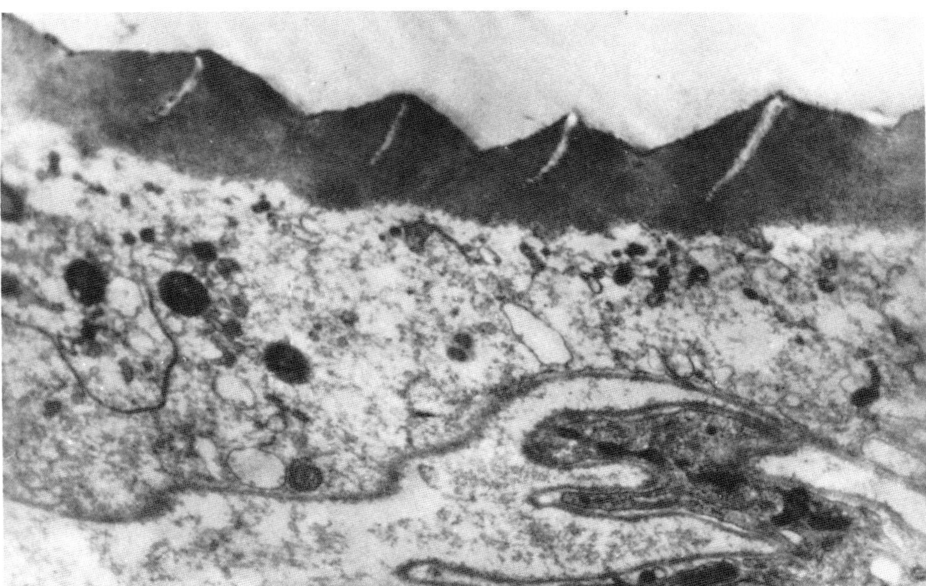

Fig. 3.41. The cytoplasmatic membrane bearing pores harmonically alternated and a substance formed by lumps can be observed, which are considered secreted material located within the cytoplasm and outside the cytoplasmatic membrane. 16940 ×

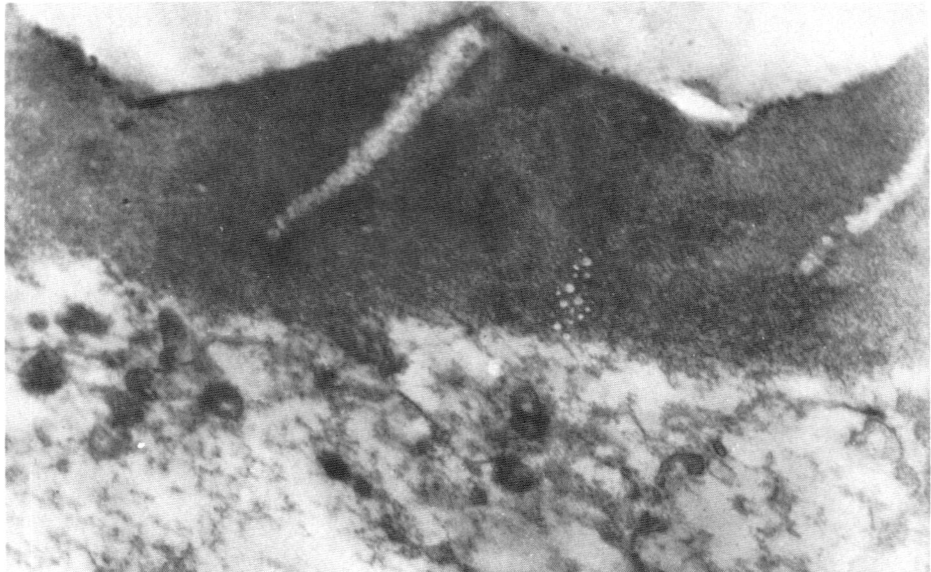

Fig. 3.42. One of the pores when eliminating the intracellular contents. (Rovere et al. 1980)

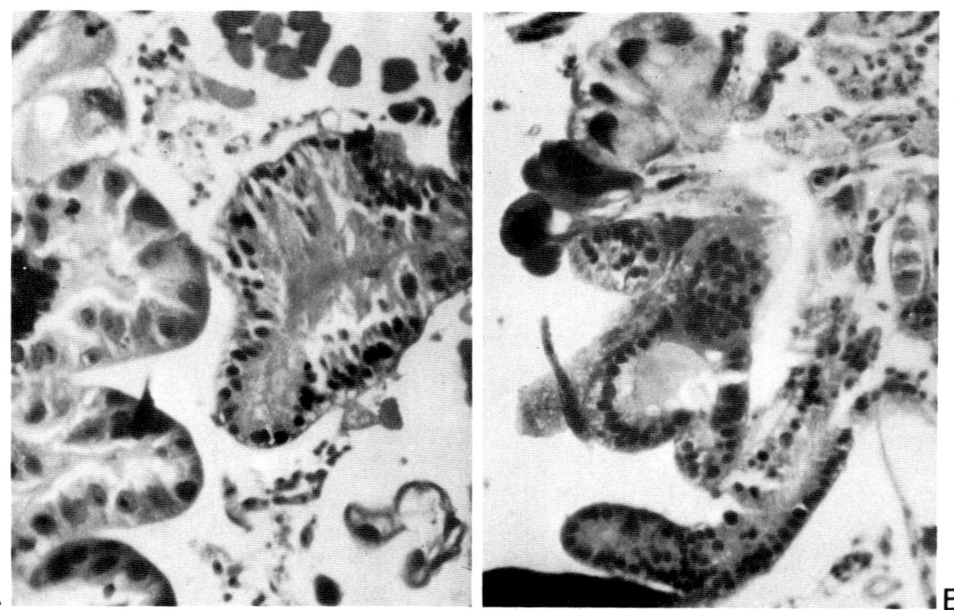

Fig. 3.43 A, B. This type of cell is located on the left lateral portion of **A**, or at the superior end, **B**

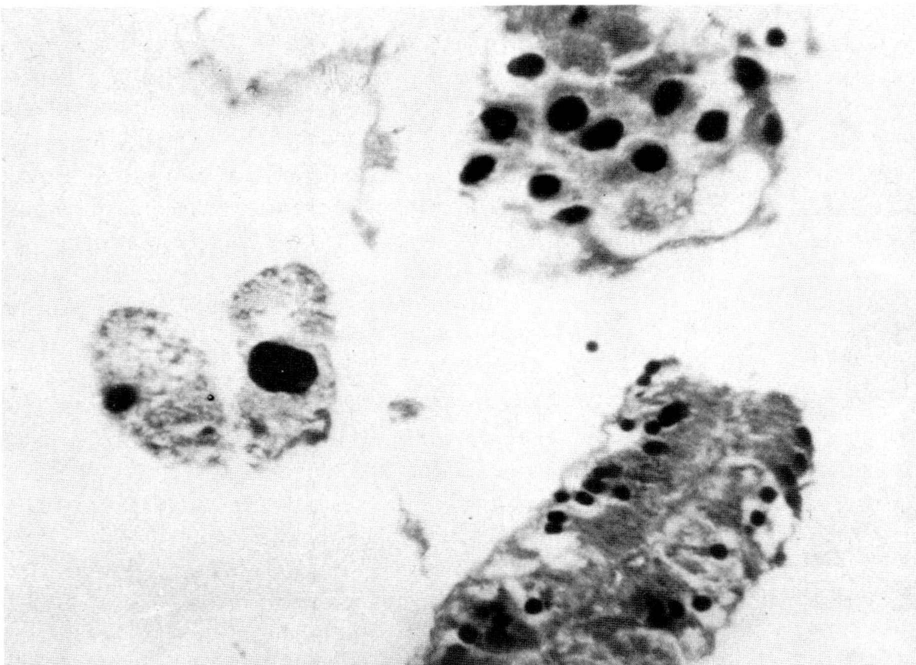

Fig. 3.44. On this photomicrography, the 3 types of cell described by Rovere et al. (1980) can be seen

been demonstrated in vivo by Tatchell (1969), and in vitro by Megaw (1976), that the salivary secretion of *B. microplus* is hypertonic with respect to the hemolymph. These glands also produce "cement", thus attaching the parasite to its host, although this has nothing to do with the damage caused to the skin of the host.

Tatchell and Binnington, in 1971, demonstrated the absence of anticoagulative and cytolytic actions in the salivary secretions of *Boophilus microplus*, in spite of the fact that a series of pharmacologically active elements were isolated, for example, prostaglandins. In fact, Higgs et al. (1976) have proven the presence of high concentrations of PGE_2 in the saliva of *Boophilus microplus*, which would play a very important role in the commencement and maintenance of damage to the host.

These authors refer to works by Moorehouse and Tatchell (1966) and by Tatchell and Moorehouse (1968) on the feeding mechanism of the tick. The buccal parts of *Boophilus microplus* are extremely short and only penetrate the skin to the malphigian layer; this would mean that the damage observed on the cattle skin beneath the epidermis neither had a traumatic origin nor was produced by lysis of the host tissues by means of the digestive action of the saliva, but that the damage was a consequence of an inflammatory reaction on the part of the host.

The saliva would determine the alterations on the level of capillaries, which would then suffer the first dermic changes; these capillaries dilated, showing edema and then haemorrhage; at the same time a simultaneous invasion of leucocytes and lymphocytes takes place in the zone. These alterations bear a deep relation to the 3 stages present during the feeding process: first, tissue fluids are taken in; then a great number of leucocytes, together with the foregoing fluids, and finally, full blood together with the two above-mentioned elements. It must be pointed out that, although the saliva bears a low percentage of proteins, as shown by Tatchell in 1969, it is, in fact, antigenically very strong; for this reason Allen et al. (1979) suggest that the antigen – antibody reaction and the activation of the complement would bear some relation to damage caused to the skin of the host.

The pharynx is a powerful suction organ. Its walls are heavily sclerotized and intimately joined to a membrane. The pharyngeal muscles are as follows: two lateral dilators, which are well developed; two dorsal dilators, smaller than the previous ones; two ventral dilators, and finally, a muscular group: the constrictor muscle, which alternates its function with that of the dilators.

A short esophagus follows the pharyngeal cavity. The esophagus bears a simple cylindrical epithelium with cells containing nitid cytoplasm and small nucleus. This epithelium becomes pseudostratified at the esophagogastric junction at the level of the cardia.

The stomach is a small sac, from which the blind guts originate towards both sides. The epithelium wrapping the organs bears cells which vary in shape and size; Arthur (1962) describes a group as follows: small cells with basal nucleus and reticulated cytoplasm, frequently containing eosinophil granules.

Likewise, the largest cells project into the digestive tract, practically occluding its lumen; a fact which becomes evident during the digestive process due to the hypertrophy of cytoplasm when blood elements are phagocytized, the nucleus keeping its original diameter.

Externally, the walls of the blind guts bear muscular fibres arranged both longitudinally and transversally. They are without doubt responsible for the ameboid movements that these elements present during the feeding period of the tick, thus allowing the movement of blood along the digestive tract, and the mixing of the blood with the digestive enzymes necessary to aid the digestion.

Caudally, the *rectal sac* follows the stomach, before opening into the anus or nephrostome.

During the feeding process of the tick, the epithelium, both of the stomach and of the diverticula, is in a state of permanent proliferation; a certain number of cells rupture, mixing with the contests, which is considerably homogenous once the breakdown of the erythrocytic membrane takes place.

Arthur (1962) mentions that, according to Hughes, the cause of the initial hemolysis may be attributed to the colloidal material present in the lumen. Hughes also suggests that the erythrocytic membrane would be one of the sources of lipoid material absorbed by the epithelium of the diverticula. These lipids would contribute to form the reserve material which the ectoparasite needs for its moulting, and also for the production of each of the eggs. On the other hand, ticks are generally believed to be able to synthesize lipoid material from the nitrogenous components that enter with the blood. In this way, it is deduced that either of these two possibilities may be the origin of the grease exuded by the tick through its cuticle while feeding.

At a certain point of time, and fundamentally due to the vermiform movements of the diverticula, the gut contents are transformed into a semi-solid mass due to loss of water. This process takes place primarily through the cuticle and secondarily through the Malphigian tubules.

At the same time, these latter structures take the metabolites which, by osmosis, have passed into the cavity of the body and which have been incorporated into the hemolymph allowing their passage, thanks to their own movements and to the contraction of the cuticular muscles, towards the rectal sac.

The rectal sac is filled with a whitish fluid of final excreta, which can be seen clearly and which, once it is discharged through the anal opening and in contact with the air, solidifies, acquiring a calcareous appearance. This material is known as guanine crystals (see Fig. 3.45).

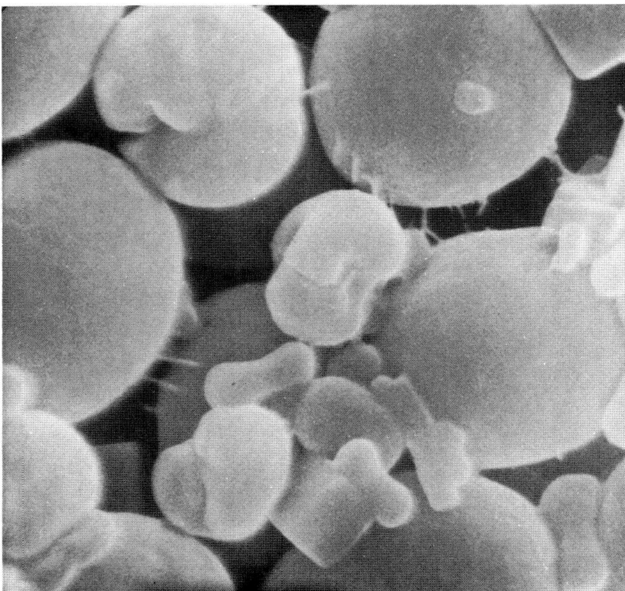

Fig. 3.45. Excreta from larva of *Boophilus microplus.* 16230×

Central Nervous System

The central nervous system is centroventrally located on the front third of the idiosoma. The esophagus crosses the brain obliquely, in a ventral dorsal direction, dividing it into two regions: pre- and postesophagal.

The brain (synganglion) is covered by a thin neurilemma and surrounded by a membrane which forms the periganglionic sinus of the dorsal aorta, and which Binnington and Tatchell (1973) represent as shown in Fig. 3.46.

These authors describe a pair of optic ganglia on the dorso-anterior portion within the pre-esophageal region of the brain; laterally and ventrally to these ganglia, and on both sides there develop the ganglia from which the chelicerae and the palps will be innervated. Dorsally from the esophagus there appears a bilateral structure, which may represent a pair of pharyngeal ganglia.

In the postesophageal region, there appear the four pairs of ganglia (1 − 4) which will innervate the corresponding four pairs of legs, and towards the caudal end, a pair of visceral ganglia (see Fig. 3.47).

The rest of the CNS structure is composed of neurons, the bodies of which constitute the cortex, whereas the neuropile is composed of their corresponding dendrites and axons (see Fig. 3.47).

Binnington and Tatchell (1973) make reference in their work to the presence within the neuropile of thicker spherical masses, which they call glomerules.

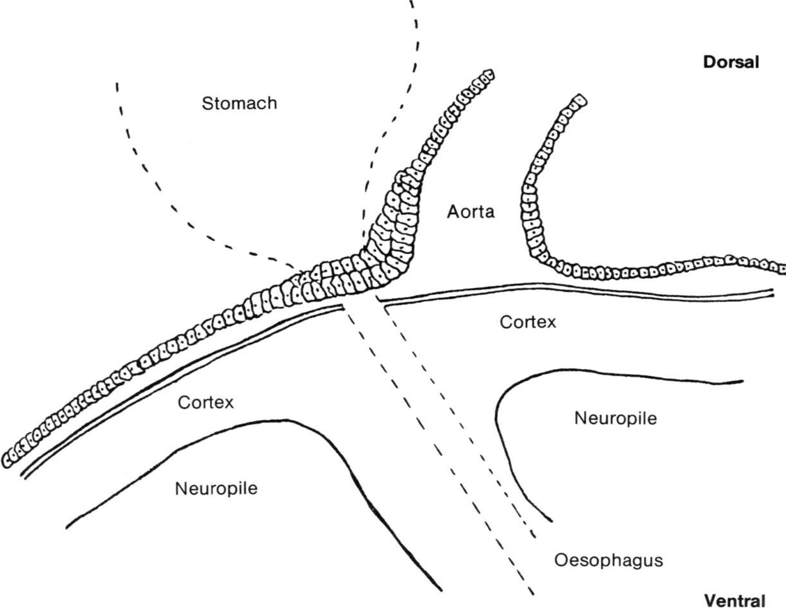

Fig. 3.46. Sagital section through the gut, aorta and dorsal region of the synganglion. (Binnington and Tatchell 1973)

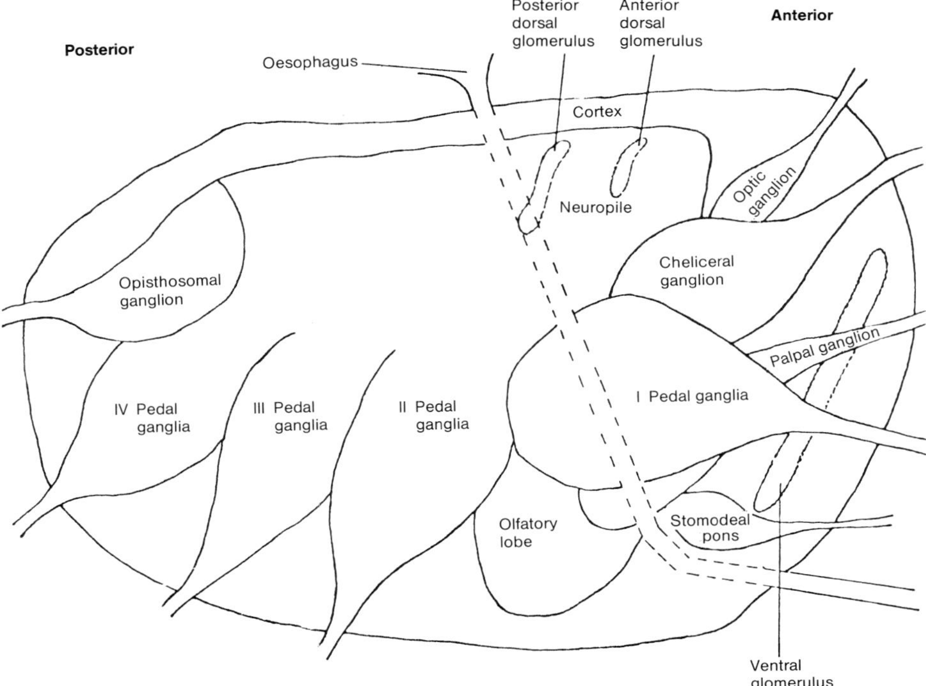

Fig. 3.47. Diagram of the brain of *Boophilus microplus*. (Binnington and Stone 1977)

They describe a pair of ventral glomerules near the ganglia of chelicerae in a dorso-anterior position with regard to the esophagus – cortex – neuropile junction. The anterior and posterior dorsal glomerules are also mentioned within the perisophageal region.

The olfactory lobules, which are highly developed, are located slightly laterally to the esophagus in the ventral region. It is observed that the peripheric nerves generally appear in pairs, with the exception of the pharyngeal and gut nerves.

There exist 15 grouped pairs of neurosecretory cells on the cortex of the brain, and in the neuropile there exist different tracts, where neurosecretions are stored.

The presence of catecholamines has been proven, not only in the CNS, but also in the peripheric nerves of *Boophilus microplus* (Binnington and Stone 1977).

Stone et al. (1978) have shown that norepinephrine is the most important in the process of neurotransmission, otherwise as neurosecretory and motor substance.

Circulatory System

By means of the circulation of the hemolymph, the products both of digestion and of excretion are distributed throughout the tick's body. Taking into account the observation of Krantz (1978) and of several other authors, this colourless substance contains three primary types of cell:

1. Minute proleucocytes with large nuclei (5 – 7.6 µm);
2. Oval basophilous haemocytes (10 – 20 µm) containing glycogen;
3. Amoeboid oesinophilous haemocytes (12 – 25 µm).

Other transitional cells can also be found, as well as amino acids, lipids and glucose.

The hemolymph circulates freely throughout the body, thanks to the pulsations of the heart, which extends from the anterior end to the posterior two-thirds of the idiosoma. As described by Arthur (1962), the heart is supported in the hemocoele by a group of extrinsic muscles; at one end two pairs of them are inserted laterally and one pair posteriorly between the ostia, all of them having their origin on the dorsal wall of the idiosoma.

Thus, diastole is achieved by the contraction of this group of muscles, whereas systole is achieved by the contraction of intrinsic musculature in the wall of the heart. As an example, Arthur (1962) mentions that the pulse rate in *Dermacentor andersoni* varies between 20 and 128 pulsations per minute, but that there is frequent cessation of heart beat for several seconds.

A non-muscular dorsal aorta leads forward from the heart, enlarging, as has already been seen, to form the periganglionic sinus (see Fig. 3.46). From here, the dorsal aorta continues as a sinus around the capitular nerves and esophagus before terminating at the posterior end of the pharynx.

Respiratory System

The *spiracular plates*, a rigid formation of the cuticle, are located posteriously to coxae IV and are oval in shape, which is characteristic of the genus *Boophilus*. These external structures are present in nymphs and in adult stages, whereas in larvae, for example, the gaseous exchange is other than cutaneous.

Eccentrically on this plate, there appears a slightly pigmented region called the macula, which in turn coincides with the external opening of the ostia.

Around the macula and on the slightly concave surface of the spiracle plate, a series of orificia of varying shape can be observed (see Fig. 3.35). These orificia, apart from being interconnected, end in a broad space located beneath the ostia, called subostial space. This space leads into a tube-like structure known as the atrium. The atrium is delimited by a dorsal and a ventral wall, and is moved by intrinsic musculature. Possibly, these movements contribute to regulate the gaseous exchange. Taking as reference the description made by Arthur (1960), the following tracheal trunks arise from the atrium and distribute to both sides of the idiosoma, namely:

1. The anterior trunk, which divides to send branches to each of the legs and to the brain.
2. The median-anterior trunk, which supplies the salivary glands and the genital system.
3. The postero-dorsal trunk, which yields tracheation for the stomach, for the posterior portion of both the salivary glands and for the genital organs.
4. Two smaller trunks, the postero-medians, which together with the postero-lateral trunk V, supply the posterior portion of the idiosoma.

Arthur (1960) believes that in general, females have a wider tracheal system than males.

Reproductive System

Male Genital Organs

The male genital organs start with paired elongated structures called testes which extend from the vicinity of the brain to approximately the posterior border of the coxa of leg IV. The vase deferentia arise from the front end of the testes, and before ending in the genital opening, are transformed into the ejaculatory ducts.

The accessory glands open precisely at this junction of the vasa differentia and the ejaculatory ducts.

According to the description made by Oliver (1974), the spermatogenesis may be divided into four periods, namely:

1. Mitosis of the spermatogonium.
2. Differentiation and increase in size of the primary spermatocytes.
3. Double meiotic division and partial development of the spermatid.
4. Differentiation and final development of the spermatid into ripe spermatozoa, after having been transferred to the female, with increase in their length and physiological activation.

Although feeding is directly important with respect to the process of spermatogenesis, it has not yet been discovered what the stimulus is. Possibly the neurosecretory cells of the brain may be involved in this process, but this has not yet been proven.

Female Genital Organs

An elongated ovary is located near the rectal sac, extending forward on both sides of the idiosoma, to continue into a pair of oviducts.

The vagina is divided into two portions: the cervical, into which both oviducts will end, and the vestibular portion which leads into the genital opening. Directly related to the cervical portion of the vagina is the seminal receptacle, and, as quoted by Arthur (1960), Douglas suggested the possibility that this sac-like structure would serve to store the spermatophore, (a group of spermatids surrounded by a capsule which the male transfers to the female during copulation).

On this same portion of the vagina and before the receptacle, a pair of accessory glands are angularly located, which help to lubricate the vagina with their contents, thus facilitating the extrusion of eggs through the vagina.

Copulation begins when the male moves beneath the female and proceeds to seek immediately the genital orifice of the female. In general, the male clings with his legs to the basal joints of the female's legs: his pair I between pairs I and II of the female and so on; resulting in the close apposition of their ventral surfaces.

As expressed by Feldman-Muhsam and Borut (1971) in some ticks of *Metastriata,* the male locates his capitulum at right angles to the genital orifice of the female, and only the chelicerae, that is without their sheaths, penetrate. These authors established the true importance of segment IV of male palps without which copulation is not possible. During copulation it was observed that males segregate saliva.

Soon after copulation, the spermatophore is extruded, the bulb of which is pushed out first and filled with a colourless fluid, as described by Arthur (1962); this is followed by the neck and capsule composed of the endo and ectospermatophores containing the spermatids, which are forcibly squeezed into the bulb by the contractions of the ejaculatory ducts.

Once the spermatophore is attached around the genital orifice of the female, the endospermatophore already in the vagina evaginates from the ectospermatophore; this remains on the surface but is detached later. If the female does not copulate, she does not complete her feeding either, for the commencement of the oviposition is directly related to the weight of the engorged female, that is she must have a minimum quantity of blood in order to start laying eggs; a fact which we have had the opportunity of proving in field trials.

As the eggs are being extruded from the vagina, they are temporarily received by the glandular organ of Gené, located beneath the dorsal shield where the excretory duct opens into the basis capitulum. This glandular organ produces a lipoid substance which agglutinates the eggs and prevents desiccation.

Cuticle (Exoskeleton)

Generally, as expressed by Arthur (1962), hard ticks have their exoskeleton formed by the following described layers, which from inner to outer are (see Fig. 3.48):

1. Epicuticle: the wax layer, the polyphenol layer and the cuticulin layer.
2. Outer endocuticle.
3. Inner endocuticle.
4. Epidermis.

The first of these is composed of a superficial wax layer, which confers impermeability to the cuticle.

The polyphenol layer is located within the epicuticle and beneath the previous one. This layer is composed of minute droplets rich in polyphenols, which then penetrate into the following layer called cuticulin. This layer consists of protein and forms a thin membrane over the surface bearing a great number of microspores, which connect with the ducts originating on the epidermis. The precur-

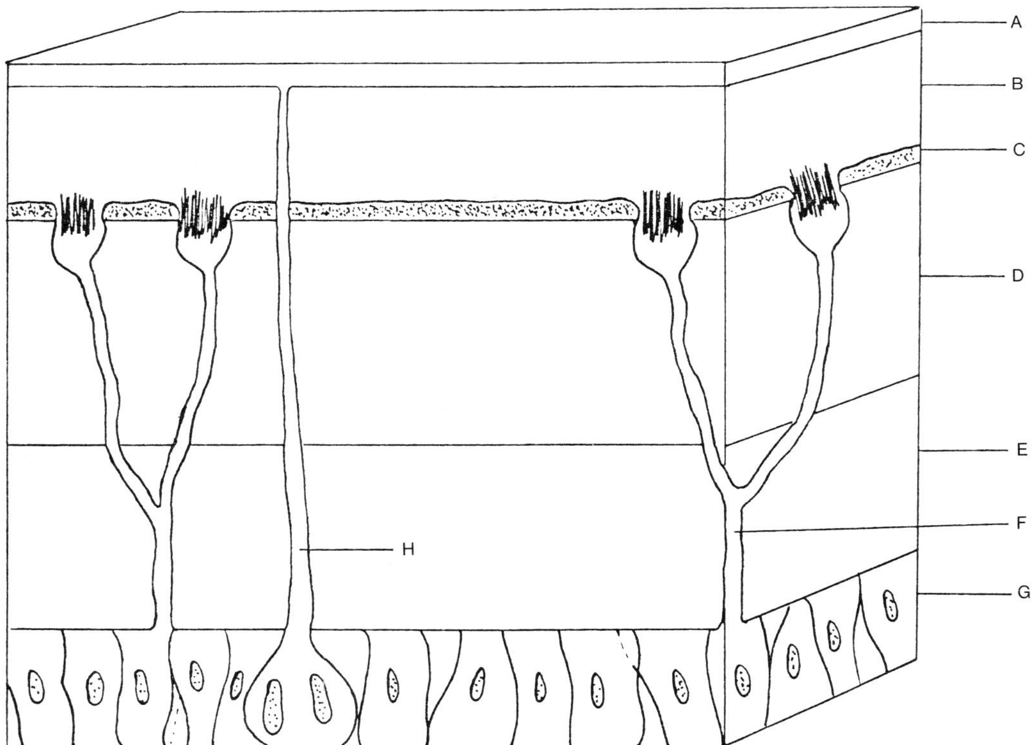

Fig. 3.48. Diagram of cuticle. *A* wax layer; *B* polyphenol; *C* cuticulin; *D* outer endocuticle; *E* inner endocuticle; *F* duct; *G* epidermis; *H* dermal gland. (After Krantz 1978)

sors of the wax layer seem to be secreted by the epidermal cells and their transport along those ducts through the cuticulin and polyphenol layers may well be facilitated by an emulsifier or a solubilizer.

Arthur (1962) describes various functions attributed to the pore canals. Apart from that already mentioned, they may transport the oxidizing agents or even the phenolic substrate and protein necessary for the sclerotization of the epicuticle.

According to Arthur, Lees showed that water evaporated from the epicuticle is replaced by water taken up by the epidermal cells from the hemocoele. Likewise, at high humidity, the water is transferred from the epicuticle and along these pore canals through the endocuticle to the epidermis, then passing into the hemocoele.

Both inner and outer endocuticle consists of laminae, chemically similar to one another, but which differ from each other in their position: in the outer endocuticle they are packed closely together and accordingly render the layer very dense, whereas in the inner endocuticle, the laminae are freely located and produce a more spongy effect. Likewise, when sections of the cuticle are stained with Mallory[1], the inner endocuticle stains blue and the outer endocuticle stains red.

The epidermis consists of cells whose function is to produce the cuticle; among these cells are larger ones, the dermal glands, which secrete some special substances. As the thick engorges, these cells hypertrophy enormously and they are linked to an exterior duct, which opens directly on the surface of the epicuticle.

References

Allen JR et al. (1979) The location of tick salivary gland antigens, complement and immunoglobulin in the skin of guinea pigs infested with *Dermacentor andersonii* larvae. Immunology 38:467 – 472

Arthur DR (1960) Ticks. A monograph of the Ixodoidea, part V. Cambridge Univ Press, Cambridge

Arthur DR (1962) Ticks and disease. Int Ser Monogr Pure Appl Biology. Pergamon Press, Oxford New York

Baker EW, Wharton GW (1952) An introduction to acarology. Macmillan, New York

Beadle DJ, Megaw MMJ (1979) Intercellular junctions in the hypodermis, salivary gland and Gene's organ of the cattle tick *Boophilus microplus*. Cell Tissue Res 202: part 1, 119 – 124

Bedford GAH (1934) South African ticks. I. Onderstepoort J Vet Sci Anim Ind 2:49

Binnington KC, Stone BF (1977) Distribution of catecholamines in the cattle tick *Boophilus microplus*. Comp Biochem Physiol 58C:21 – 28

Binnington KC, Tatchell RJ (1973) The nervous system and neurosecretory cells of *B. microplus* (*Acarina, Ixodidae*). Z Wiss Zool 185: no 3/4, 193 – 206

---

1 [basic acid fuchsin, orange G and blue anilin. Lillie (1954), Rameis (1928)]

Boero JJ (1944) Los ixodídeos de la República Argentina. Boletín Tecnico de la Dirección General de Ganadería. Minst Agric Dir Sanidad Anim Buenos Aires

Boero JJ (1957) Las garrapatas de la República Argentina (*Acarina Ixodidae*). Univ Buenos Aires, Dep Ed

Chow YS, Lin SH, Su JS (1972) A new tarsal gland of the brown dog tick *Rhipicephalus sanguineus* (Latreille), 1804 (*Acarina, Ixodidae*). Bull Inst Zool Acad Sin 11:35 − 39

Cooley RA (1946) The genera *Boophilus, Rhipicephalus* and *Haemaphysalis* (*Ixodidae*) of the new world. Nat Inst Health Bull 187. USA Government Printing Office, Washington DC

Corwin D, Clifford CM, Keirans JE (1979) An improved method for cleaning and preparing ticks for examination with the scanning electron microscope. J Med Entomol 16: no 4, 352 − 353

Feldman-Muhsam B, Borut S (1971) Copulation in ixodic ticks. J Parasitol 57: no 3, 630 − 634

Fielding JW (1926) Australasian ticks. Commonwealth Aust Dep Health, Serv Publ (Trop Div) no 9

Gothe R (1967) Ticks in the South African Zoological Survey Collection. Part XII The general *Boophilus*. Curtice 1891 and Margaropus Karsch, 1879. Onderstepoort J Vet Res 34:81 − 108

Haller G (1881) Vorläufige Bemerkungen über des Georganode. Isodiden Zool Auz 4:165 − 167

Higgs GA et al. (1976) Prostaglandins in the saliva of the cattle tick *Boophilus microplus* (Canestrini) (*Acarina, Ixodidae*). Bull Entomol Res 66:665 − 670

Hoogstraal H (1956) African Ixodidea. Ticks of the Sudan (with special reference to Equatoria Province and with preliminary reviews of the genera *Boophilus, Margaropus* and *Hyalomma*). Dept Navy Bur Mod Surg. Washington DC 1101 pp

Keirans JE, Clifford CM, Corwin D (1976) Ixodes Sigelos, N.S.P. (*Acarina: Ixodidae*), a parasite of rodents in Chile, with a method for preparing ticks for examination by scanning electron microscopy. Acarología T XVIII, Fasc 2

Krantz GW (1978) A manual of acarology, Second edition. Oregon State Univ Book Stores, Corvallis, Oregon

Lombardero OJ (1971) Glosario de términos parasitológicos. Ed Univ Buenos Aires

Londt JGH, Arthur DR (1975) The structure and parasitic life cycle of *Boophilus microplus* (Canestrini 1888) in South Africa (*Acarina Ixodidae*). J Entomol Soc South Afr 38 (2):321 − 340

McMullen HL, Essenberg RC, Saver JR (1979) Fed Proc 38/3.

McMullen HL, Essenberg RC, Saver JR (1979) Modulators of phosphodiesterase and the process of fluid secretion in the salivary glands of an ixodid tick. Oklahoma State University Siillwater. OK Fed Proc 38 − 3

Megaw MJW (1976) Structure and function of the salivary gland of the tick *Boophilus microplus*. Ph D Thesis. Univ Cambridge, England

Megaw MJW, Beadle DJ (1979) Structure and function of the salivary glands of the tick *B. microplus* Canestrini (*Acarina: Ixodidae*). Int J Insect Morphol Embryol 8 (2):67 − 83

Moorehouse DE, Tatchell RJ (1966) The feeding process of the cattle tick *Boophilus microplus* Canestrini. A study in host − parasite relations. I. Attachment to the host. Parasitology 56:623 − 632

Nuñez JL, Pugliese ME, Hayes RP (1972) *Boophilus microplus* Can. Estudios sobre los estadios parasitarios del ciclo biológico. Rev Med Vet Buenos Aires, 53 (1):19 − 35

Oliver JH Jr (1974) Symposium on reproduction of arthropods of medical and veterinary importance. IV. Reproduction in ticks. (Ixodoidea). J Med Entomol 11: no 1: 26 − 34

Roberts FHS (1969) The larvae of Australian Ixodidae (*Acarina: Ixodidae*) J Aust Entomol Soc 8:37 − 78

Rovere RJ, Borondi A, Perez Arrieta A, Marti Vidal J, Krivoruchky I (1980) Estudio cito-histológico experimental óptico y electrónico de la estructura funcional de glándulas salivales de garrapatas. III Congr Argent Cienc Vet, Buenos Aires

Rudolph D, Knulle W (1974) Site and mechanism of water vapour uptake from the atmosphere in ixodid ticks. Nature (Lond) 249:84–85

Slifer EH (1970) The structure of arthopod chemoreceptors. A Rev Ent 15:121–142

Stone BF, Binnington KC, Neish AL (1978) Norepinephrine as principal catecholamine in a specific neurone of an invertebrate (*Boophilus microplus: Acarina*). Experientia 34:1173–1174

Tatchell RH (1969) The ionic regulatory role of the salivary secretion in the cattle tick *Boophilus microplus*. J Insect Physiol 15:1421–1430

Tatchell RH, Binnington KC (1971) An active constituent of the saliva of the cattle tick *Boophilus microplus*. Proc Int Congr Acarol (Prague) 1971:745–748

Theiler G (1943) Notes on the ticks of domestic shock from Portuguese East Africa. Estaçao Anti-Málarica de Lorenço Marques. 55 pp

Tsvileneva VA (1964) The nervous structure of the ixodid ganglion. Zool Jahrb Abt Anat 81:579–602

Waladde SM (1976) The sensory nervous system of the adult cattle tick *Boophilus microplus* (Canestrini) Ixodidae, part I. Light microscopy. J Aust Entomol Soc 15:379–387

Waladde SM (1977) The sensory nervous system of the adult cattle tick *Boophilus microplus* (Canestrini) Ixodidae, part II. Scanning electron microscopy. J Aust Entomol Soc 16:73–79

IV Life Cycle

Every action, whether defensive or offensive, directed towards the control or eradication of the common cattle tick must be based upon a deep knowledge of its life cycle, not only of the mean data but also of the extreme variations corresponding to each stage. For this reason, several authors from different countries have studied the life cycle, both of the free-living and the parasitic forms, and given us very complete information.

Given the importance of this aspect, it will be studied here in detail.

Free-Living Cycle

Once the engorged or ripe ovigerous female drops off the host to the ground, her free-living cycle starts. This can be divided into the following periods:

- Preoviposition
- Oviposition
- Postoviposition
- Incubation
- Eclosion
- Free-larval life.

Preoviposition

When the fully engorged female spontaneously drops off the host, she wanders in search of shadowy and protected sites for egg-laying. The period between dropping off and the beginning of oviposition is known as preoviposition.

According to Lahille, who has studied in detail the free-larval cycle of *Boophilus microplus* in Argentina, preoviposition lasts from 2 to 4 days in summer; in winter, this range extends to 90 – 97 days. The mean data achieved throughout one year of observation are described in Fig. 4.1 (Lahille 1917).

Ivancovich (1975), in Formosa, Argentina, gives data ranging from 2.3 to 2.5 days between December and February, and a maximum of 6.7 days in the month of July.

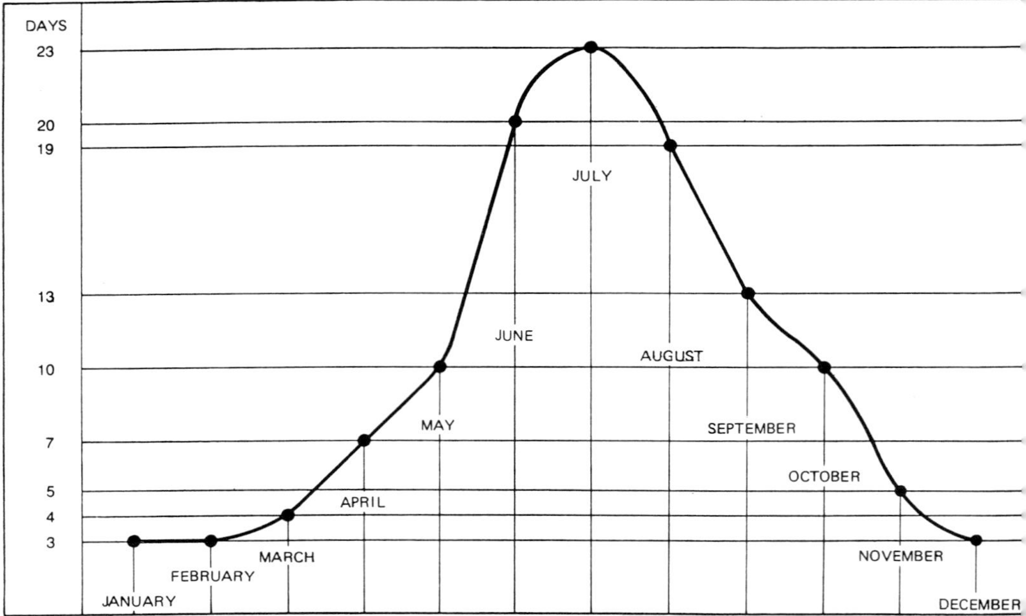

Fig. 4.1. Mean preoviposition

In over a total of approximately 6000 observations of our own, the data obtained in summer ranged from 2 to 6 days, with a mode of 3 and an average of 3.4 days. In general, 95% of the preoviposition periods lasted from 2 to 4 days, in the summer months under laboratory conditions. In other countries such as India, Sapre (1940) gives data of periods of 7 days with an average of approximately 4.2.

Legg, in Australia (1930), found figures ranging between 2 and 12 days, giving the following seasonal variations: winter 5 − 9 days, autumn and spring 4 − 6 days, summer 2 − 4 days.

In a recently published work two Brazilian investigators, Alvarado and Gonzalez (1979), give data of average 3 days (86% of the engorged females studied) with a range of 2 to 6 days.

In yet another work published in 1980, Davey et al., in the USA, give an average of 3 − 3.2 days under laboratory conditions at 27 °C and 80% relative humidity.

Oviposition

This is the name given to the period extending from the laying of the first egg by the engorged female to that of the last one.

Environmental factors have a definite influence upon the duration of this period. Thus Lahille gives a period of 13 to 15 days, as the normal in summer, and of 35 to 45 days in winter; a minimum period of 8 days in summer (January) and a maximum of 60 days in winter (August). These data were obtained in Buenos Aires. Ivancovich, in Formosa, has observed a period of between 9 and 10 days in December and January, and of 27 days in August.

Sapre, in India, gives an average of approximately 18 days, and 14 − 24 days as the range.

On the other hand, Legg, in Australia, gives a period of 5 days in summer and 30 in winter.

In general terms it can be said that oviposition lasts twice as long in winter as in summer, and, under severe climatic conditions, it can be three times as long in winter.

With regard to the number of eggs laid by an engorged female, we will present a compilation of data published by different authors, because the very different methods used for counting the number of eggs can greatly affect the results which are obtained.

Sapre, in India, gives data obtained during the month of October in the Northern Hemisphere, where out of a total of 16, 25 and 41 engorged females, he observed averages of 2338; 2300 and 2451 eggs per parasite. In this particular case, the counting system designed by Nutall was applied: the direct count on a slide divided into squares.

Legg, in Australia, reports data compiled after observing 120 ovigerous females: a maximum of 4269; a minimum of 1673; an average of 2579. The greatest number of eggs laid by an engorged female during a 24-h period was 731.

Lahille, in Argentina, reports an average of 3000 eggs per specimen, ranging from a maximum of 4502 to a minimum of 1462.

Quevedo et al. at the *II Congreso Nacional de Veterinaria, Buenos Aires* (1960), published a counting method based upon the weight of the oviposition: taking into account the average of several weights of batches of 10 eggs each, they reached the conclusion that the weight per unit was of 44 gammas (millionth grams), from this they estimated that each gamma of a normal oviposition contains 22,727 eggs.

Considering the weight of eggs, works by Davey et al. give an average of 47.7 µg at oviposition. These data coincide with those obtained by Quevedo et al. in Buenos Aires in 1960. Davey et al. recorded a remarkably high oviposition: 5465 eggs per engorged female.

Likewise, Ivancovich, using the previous method, gives a range of data of 2014 and 2134 in December and January (minimum), and of 2855 in June (maximum), giving a general average of 2391 per adult female.

Alvarado and Gonzalez recorded an average oviposition of 3285 eggs per engorged female in Brazil.

In our own normal practice, we use both systems, and have shown great variations: in brief, after having studied the oviposition of 4500 engorged females, we observed that the mean oviposition of a *Boophilus microplus* engorged female ranges between 2500 and 3000 eggs, with a range of between approximately 1000 and 4500.

Postoviposition

This is the period of time between the finish of oviposition and the death of the ovigerous female, after having fulfilled her role.

According to our results, which include the above-mentioned cases, this period lasts from 2 to 15 days. The number of ticks which survive after 8 days is exceptionally low. In these latter cases, slight movement occurring at long intervals can be observed only by means of prolonged observation, especially of the dorsal region.

According to Legg, the death of the ovigerous female that has already fulfilled her role occurs very near to the end of the oviposition in summer, whereas in winter this time is prolonged, a fact which has also been observed within the Argentine Republic.

Incubation

Once the eggs have been deposited in the environment they present an elliptical shape and are of approximately 550×400 μm. They are dark brown in colour, having a sticky and glistening surface covered by an albumin-like substance. During the first half of the incubation period, only a few details can be observed on the embryo, whereas in its second half, the excretory canals and the three pairs of legs begin to differentiate. This is the period of the life cycle when environmental factors may most influence the evolution of the embryo; for temperature and humidity conditions, especially the former, can considerably shorten or prolong the incubation period.

For example, Legg, in Townsville, Queensland, where winters are not extremely severe, gives the following data: minimum period of 15 days in summer and maximum of 51 in winter; whereas Lahille, in Argentina, gives data ranging from 19 to 51 days with an average of 22. These data coincide with those we have obtained, although the vast majority of our data arise from cultures kept in vitro, either in glass vessels (Petri dishes or plates) or in Metianiu tubes (large test tubes) where the variations of temperature have always been in some way attenuated. Ivancovich (1975) gives an average time of 17.3 days in December (range $14-29$) and an average of 68.3 in June (range $58-75$).

Hooker et al. (1912) report a period of 24 days under laboratory conditions, but a range of 27 to 34 days under natural conditions. This conclusively shows that under field conditions the incubation period is longer than that observed in the laboratory.

Alvarado and Gonzalez observed, again under laboratory conditions (26 °C and 80% relative humidity), that, for 77% of the batches of eggs deposited, incubation lasts from 22 to 24 days, with a range of 21 to 27.

Eclosion

Percentages of eclosion under laboratory conditions are quite high, always above 80% provided that the specimens are normal, that is, without visible morphological alterations and without having suffered from handling or having been affected by any ixodicide treatment, or subjected to excessive heat, sunlight, etc.

Legg studied 127 batches of eggs during a 6-month period; 57 of them had a fertility percentage higher than 90%, 50 had a percentage between 80 and 90%, and the remaining batch of 10 had 68%. Lahille reports data ranging from 60 to 98% in summer.

Oviposition may be seriously affected detrimentally by direct sunlight in the environment. Even during periods as short as an hour at any stage of the incubation period, sunlight practically destroys all of the eggs. Even on cloudy days, only a few hours are needed for their total destruction.

Another important factor as regards fertility is that of relative humidity: although percentages from 80 to 90% notably favour fertility, an excess, especially a permanent one, has very detrimental consequences: in vitro, humidity favours the development of fungi which, once they have attacked a culture, sterilize it; moreover, under field conditions it has been shown that the fertility index of eggs of *Boophilus microplus* submerged in water from 7 to 14 days, goes down to minimum and as time goes by, the fungi reduce the index to zero. That is why in flooded grazing fields the number of larvae is frequently much lower than in drier fields. Although sporadic floods do favour the transport of eggs over great distances, with resulting spread of the infestation, they also destroy a high proportion of the potential infestation from the affected lands.

Free-Larval Life

The larva of *Boophilus microplus* is approximately 500 μm long and 400 μm wide. It has a slightly ovoid shape and bears three pairs of legs. At the beginning it is very light amber in colour, but later it turns dark reddish. The dorsal shield covers about two-thirds of the total length of the body, and inside it, the blind guts are clearly visible; they are slightly opaque and brown in colour when they are empty.

Ventrally the rectal sac, bearing white guanine crystals, can also be clearly seen.

Shortly after eclosion the larvae climb to the top of the grass stalks, where they locate in great numbers, preferring the shady side of the fibre; they move throughout the course of the day so as to avoid sunlight if possible. These groups of larvae notably increase their activity when they detect the movement of a body in the vicinity, adopting a curious position: they steady themselves on their two posterior pairs of legs and extend their anterior pair, trying to attach themselves to the possible host.

The duration of the free-larval life is of great importance when attempting control and especially eradication of the tick; for this reason this subject has been meticulously studied in several countries. In Argentina, Ivancovich (1975) conducted an excellent study which, in our opinion, has not received the widespread distribution it deserved. Ivancovich represented graphically (Fig. 4.2) the duration of the non-parasitic stage of the life cycle, demonstrating the different

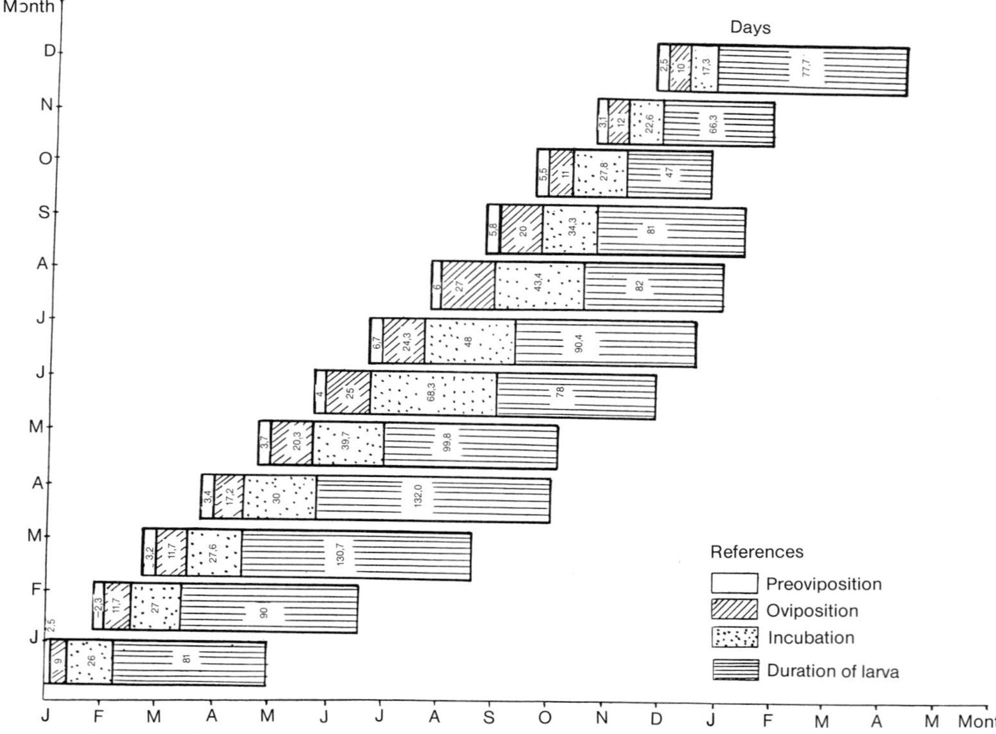

Fig. 4.2. Duration of stages of the non-parasitic phase

lengths of preoviposition, oviposition, incubation and larval life, month by month, throughout a year.

Important and interesting conclusions can be drawn from this graph, including the influence exercised by ecological factors upon the free-living cycle, the effect of the movement of herds and ixodicide treatments, and the control possibilities offered by its knowledge.

We consider it proper to include a table reproduced from the above work where the results of the study of the free-living cycle obtained under laboratory and field conditions are described (see Table 4.1).

Lahille gives the following times for considering that fields have been "cleared" of common tick larvae: *summer*: from December 15th, 65 days; *autumn:* from April, 200 days. From studies made in vitro, he gives data ranging from 10 to 70 days in summer and 250 days in autumn, winter and spring.

We have observed that light, both solar and to a lesser extent artificial (which markedly excites larvae), as well as high temperatures and low relative humidity, are factors which notably shorten the free-life of larvae. In some cases we observed a longevity of no longer than 4 days within a group of several thousands of larvae permanently subjected to these severe conditions; but on the other hand when the cultures were kept in the shade and the average temperature was $20-22\,^{\circ}\mathrm{C}$ and 80% relative humidity, we observed that larvae survived up to 204 days; these trials were carried out in vitro, using Metianiu tubes.

Table 4.1. Comparison of results obtained in the laboratory and in the field, per month of detachment of engorged females

Month	Pre-oviposition		Oviposition		Incubation		Larvae duration	
	Laboratory Days	Field Days	Laboratory Days	Field Days	Laboratory Days	Field Days	Laboratory Days	Field Days
January	3	3	9	9	–	–	–	–
February	2	3	10	9	–	–	–	–
March	3	3	10	10	29	30	130	131
April	4.3	6.6	20.6	20.3	32	32.5	124.5	133.2
May	5	5	20	21	40	49	105	107
June	4	6	33	32	70	75	56	55
July								
August	6	6	27	24	45	34	83	55
September								
October	5.5	6.5	11	11.5	–	–	–	–
November	3.5	4.5	12	12.5	25	23	60	63
Average	4.1	5.2	16.7	16.5	36.9	38.3	106.3	108.6

Average in laboratory = 164.0 days
Average in field   = 168.6 days

Legg, in Australia, worked with 150 groups of several hundreds of larvae each, and observed a maximum longevity of 150 days in winter, the maximum data reaching 115 days, in the shade and with a high relative humidity.

Incidence of Environmental Factors Upon the Development of the Non-Parasitic Life Cycle

From all the facts detailed it can be seen that environmental factors have a remarkable influence upon the free-living cycle. In suitable conditions, for example heavy rains, high grass stalks (which provide plentiful shade) and high temperatures which shorten the incubation period the most, the whole non-parasitic phase may be less than 3 weeks.

On the other hand, cold weather (which notably prolongs the incubation period), and poor rains (which diminish the quantity of grass, thus reducing the chances of ovigerous females and larvae finding shady sites) are adverse factors for the survival of *Boophilus microplus.*

Parasitic Life Cycle

Once the larva finds a preferred cattle host the parasitic cycle starts. This cycle is characterized by little variation in duration.

Some authors, among them Legg, studied this cycle in winter and in summer, but found no significant differences (see Table 4.3). Likewise Sapre (1940), in India, worked in places where the severe climate allows only two generations of *Boophilus microplus* to develop per year. He confirmed the results, which also agree with what Hitchcock (1954) had observed in Yeerongpilly, Queensland, Australia.

Given the relative lack of variation of the parasitic cycle, this allows us to structure control systems confidently, to prove the efficacy of ixodicides, and to set appropriate intervals between treatments.

The parasitic cycle in general may be divided into the following stages: larval, nymphal and adult, which in turn comprise the following instars, namely:

Larval stage: *Three pairs of legs*
                Hypostome: Double row of teeth
                Neolarva
                Larva type A
                Larva type B
                Larva type C
                Engorged larva

Nymphal stage: Four pairs of legs
Hypostome: Double row of 3/3 teeth
Young male
Adult male
Undistended female
Semi-engorged female
Engorged female

Larval Stage

Its most important morphological characteristics are: three pairs of legs and a double row of teeth in the hypostome.

Neolarva

Its morphology is similar to that of the free-living larva already described. Once the larvae are on the host they wander freely over it in search of appropriate sites to which to attach, in general those zones of tender skin rich in vascularization, such as the inner side of the thighs, perineal region, dewlap, neck and anterior border of the ears. Generally 95% of the parasitic population of *Boophilus microplus* are found in these regions.

The vast majority of larvae attach within a few minutes, and after 24 h 90% of them have already started feeding (Lahille 1917). In an experimental infestation carried out by Nuñez et al. (1972) with 30,000 larvae on three different animals, it was proved that after 24 h only 7.5% remained at the neolarval instar; the rest of them (92.5%) had already attached themselves to the host.

Larva Type A

It is this larva which, after making contact with the host, perforates the skin with the chelicerae, fixes the hypostome and starts feeding. This larva is approximately the same size as the neolarva, that is from 0.60 to 0.66 mm long and from 0.40 to 0.43 mm wide behind the dorsal shield. Remnants of host tissue can be seen in the larva's buccal apparatus. The three pairs of legs are readily visible, in turn losing mobility as the parasite starts feeding.

The division of parasitic larvae into types A, B and C dates from 1970 and was published in 1972 by Nuñez et al.; this division is based on the necessity of adopting a rational criterion for the measurement of the residual effect of an ixodicide. This is of fundamental importance in Argentina where the products are used for eradication, and hence for the total "cleansing" of animals so as to

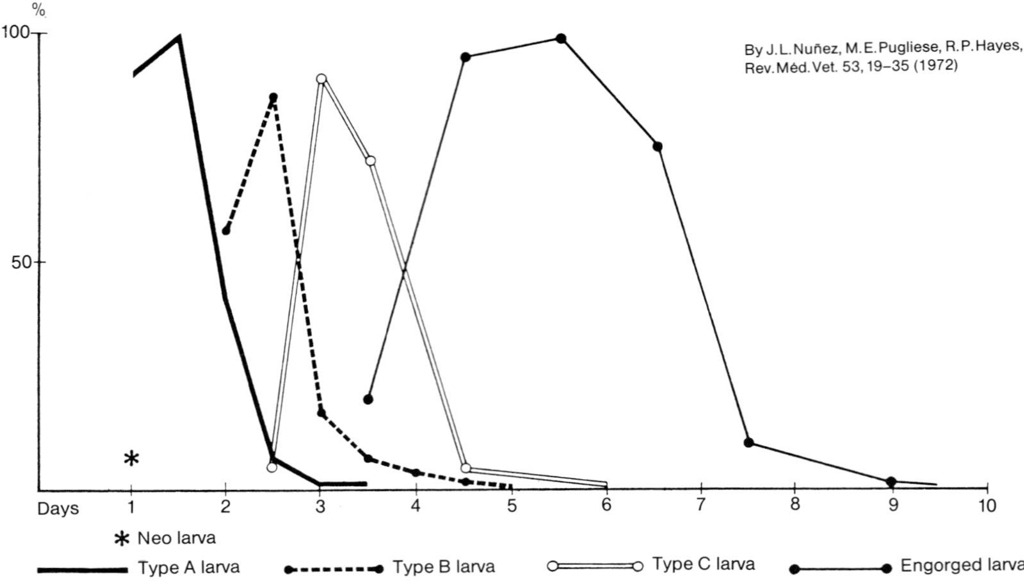

Fig. 4.3. Larval stages (Nuñez et al. 1972)

send them later to tick-free areas, must have an effective residual action. In general terms, type A, B, and C coincide with those visible instars 24, 48 and 72 h after infestation, but this cannot be taken as absolute, as considerable variations exist, as can be observed in Fig. 4.3 on larval instars (see also Fig. 4.5).

In larva type A, the shield covers two-thirds of the total length of the body, and this is the most noticeable characteristic for its differentiation. On studying the development of several thousand larvae the following conclusion was drawn: 24 h after the infestation there is a net predominance of larvae type A, and a very small percentage (less than 1% of the total number), survive up to 2 to 3 1/2 days later.

Larva Type B

Instar which chronologically follows the previous one. Here, a considerable increase in the size of the body is observed, and dorsally the shield covers approximately just less than half the total length of the body; its longitudinal axis is never larger than that of the notum, but is not less than a quarter of the total length of the body either. The legs are capable of movement, but it is not as active as the neolarva or larva type A. The colour of this type of larva tends to become lighter, turning from dark red to a more yellowish tone.

This larva, wrongly called 48-h larva, may also be found after 72 h (17.5%), after 96 h (4.5%) and an extremely low percentage (0.55 and 0.25%) can also be found 108 and 120 h after infestation (Nuñez et al. 1972).

In other words, although the majority are found between 48 and 72 h (86.25%), it is necessary to take into account the variations detailed above when a thorough examination is being carried out.

Larva Type C

This is also called 72-h larva. The longitudinal axis of the shield is smaller than a quarter of the total length of the body. The movement of the legs weakens and gradually diminishes on the distal joints until, shortly before complete immobility (engorged larval instar), only movements at the coxofemoral joint occur.

The colour of the larva becomes reddish-yellow, tending to white.

In work done in Argentina, Nuñez et al. (1972) recorded the first appearance of larva type C 2 1/2 days (60 h) after infestation, the majority 82.5% were found at 72 h; some specimens survived between 5 1/2 and 6 days after infestation (1.33 and 0.58% respectively).

Engorged Larva

It has been already noted that larva type C shows a visible decrease in the movement of its legs until complete immobility is reached. Here, the engorged larval instar begins; first the metamorphotic stage, followed by the nymphal period.

The engorged female is about 1 mm long (1.15 − 0.75), its integument distends and is creamy white in colour.

Some authors, e.g., Boero and D'Angelo (1946), give a series of morphological details, namely: the presence of the buccal extremity of the blind guts beneath the shield; others, such as Hitchcock (1954), give importance to the turgescence of the cuticle against a certain pressure. According to our observations these details, while of some interest, do not occur so constantly as to consider them absolute.

As time goes by and moulting approaches, the size of the parasite increases, reaching a total length of approximately 2 mm. In our own study of the parasitic cycle of *Boophilus microplus*, we observed the first engorged larva after 3.5 days, the highest percentage after 5.5, and the last larva after 9.5 days. Londt and Arthur (1975), South African authors, made an interesting contribution to the morphological study of *Boophilus microplus*. They studied the weight of the different larval instars: the mean weight of neolarvae is 0.031 mg, decreasing to 0.030 mg after the first 24 h. After 48 h, the weight increases to 0.038 mg, then reaches 0.065 mg after 72 h and 0.183 mg after 96 h. Finally, on the 5th day, the time of engorged larval predominance, the average weight is 0.236 mg.

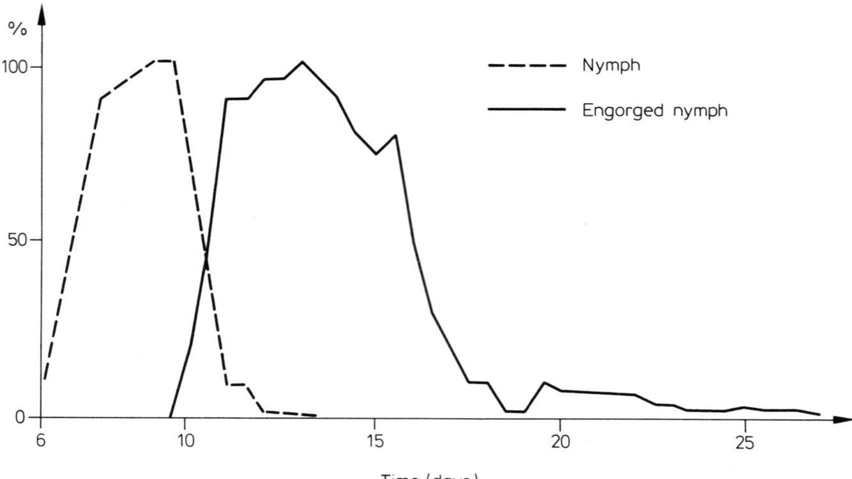

Fig. 4.4. Nymphal stages (Nuñez et al. 1972)

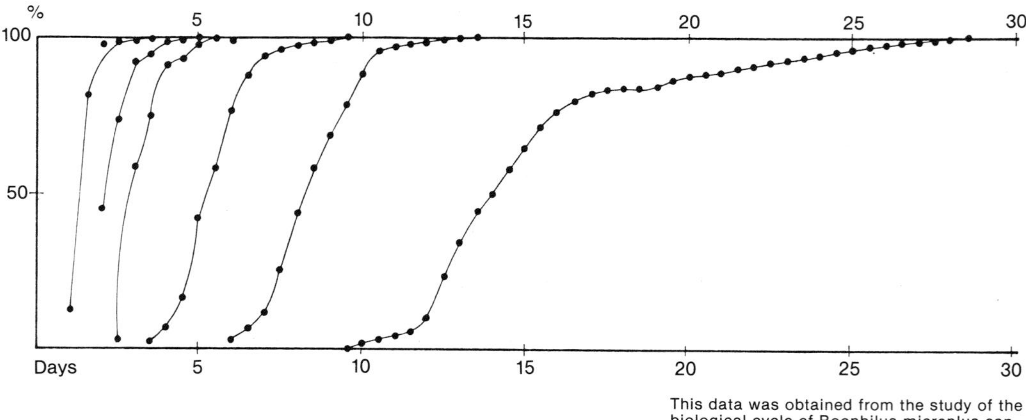

Fig. 4.5. Projection of the larval and nymphal stages of *Boophilus microplus can*

These data give not only an idea of the hematophagous capacity but also of the quick development during the first days of the parasitic cycle.

Nymphal Stage

During this stage the most outstanding morphological characteristics are as follows: 4 pairs of legs and double row of 3/3 teeth in the hypostome. The spiracles also show on both sides of the body behind leg IV (Figs. 4.4 and 4.5).

Nymph

During the first hours of the stage described above, that is engorged larval stage, beneath the cuticle there can be seen the new form which will emerge and attach itself again on to the skin of the host: this is the nymph, which will appear after the posterior part of the integument breaks. The remnants of this integument (exuviae) remain fixed to the epidermis.

The new parasitic form is, at the beginning, smaller than the previous one. It measures more than 1 mm long, and shows a translucent hyaline colour, which allows the blind guts, the rectal sac and the excretory ducts to be clearly seen.

The nymph does not wander far, and generally attaches itself again to the host and begins engorging nearly beside and in front of its previous location, becoming greyish in colour. Gradually the legs lose mobility: first, as always, on the inferior joints, which eventually become rigid, thus moulting to the following instar, called the engorged nymph. In accordance with our observations (Nuñez et al. 1972), the first nymphs appear 6 days after infestation, and by 9 days the majority had appeared (98% of the whole number of parasites) with the last ones after 13.5 days, but always a very low percentage: 0.75%.

Engorged Nymph

Once the nymph becomes immobilized, it metamorphoses to the engorged nymph. This instar, within the parasite cycle, is quite long.

The engorged nymph is of an oval shape, of a greyish brown colour, and measures about 2.5 mm at the beginning and reaching nearly 4 mm at the end of this stage. The body notably narrows behind the last pair of legs, at the level of the peritremes and, during the last days of this stage the sexual dimorphism is well differentiated: the engorged nymphs, from which undistended females are produced, are larger and lighter in colour; whereas those smaller and darker in colour will become young males. The former weigh about twice as much as the latter.

According to our observations, the first engorged nymph appeared 9 1/2 days after infestation, and at 13 days we found the highest prevalence (more than 99% of parasites were engorged nymphs) with the last ones to appear (always as a low percentage) about 28 1/2 days. In these cases the parasites were located on the anterior border of the ears.

Legg, who studied in detail the parasitic cycle, describes cases where engorged nymphs are present 21 days after infestation; and Hitchcock (1954), in an excellent publication on this subject, gives 20.3 days as the time the last engorged nymph appeared. In both cases the authors do not give information about the location. Londt and Arthur (1975) also studied this last stage and recorded the mean weight of numerous specimens, giving the following data: from the 6th until the 9th day, the weight increases insignificantly. On the 10th and 11th days, the engorged nymphs double their previous weight and, on the 14th day, they reach the maximum weight observed, 1930 mg.

Adult Stage

The most outstanding morphological characteristics of this stage are as follows: four pairs of legs and a double row of 4/4 teeth in the hypostome.

Male

When the integument of the smaller and darker engorged nymphs opens longitudinally, the males emerge. At the beginning they are rather translucent, of a greyish colour, after some hours becoming dark brown.

The cephalic end and the legs are light brown, tending to yellowish.

The total length of the body ranges between 2 and 2.5 mm and its width from 1.15 to 1.30 mm. Its 8 legs are relatively strong, with great mobility.

Ventrally, the genital orifice is seen at the level of leg II and, on the posterior third of the body the anus or nephrostome can be seen between the two pairs of adanal plates.

From a dorsal or ventral position the chitinous spur is caudally clearly noticed, and this gives an unmistakable characteristic to this stage.

Once ecdysis has taken place, the male feeds and then, taking advantage of his capacity for attaching and detaching at will, he either looks for a female to mate with or locates behind an engorged nymph, which will give rise to undistended females.

We observed the first male 13.5 days after infestation; 42 days later males constituted 100% of the parasitic population; females, once they have fulfilled their cycle, drop off the host (Nuñez et al. 1972).

Legg (1930), in Australia, quotes observations up to 46 days after infestation.

Undistended Female

The bigger and heavier engorged nymphs, generally 50% of the population, become pubescent females called undistended females. At the level of the spiracles they measure, approximately 2 mm long (extreme sizes 1.9 – 3.1 mm) × 1.3 mm wide (1.1 – 1.6 mm). Their oval, flat body is light brown at the beginning, becoming darker; their 8 legs are long and strong. Ventrally the genital orifice appears at the level of leg II.

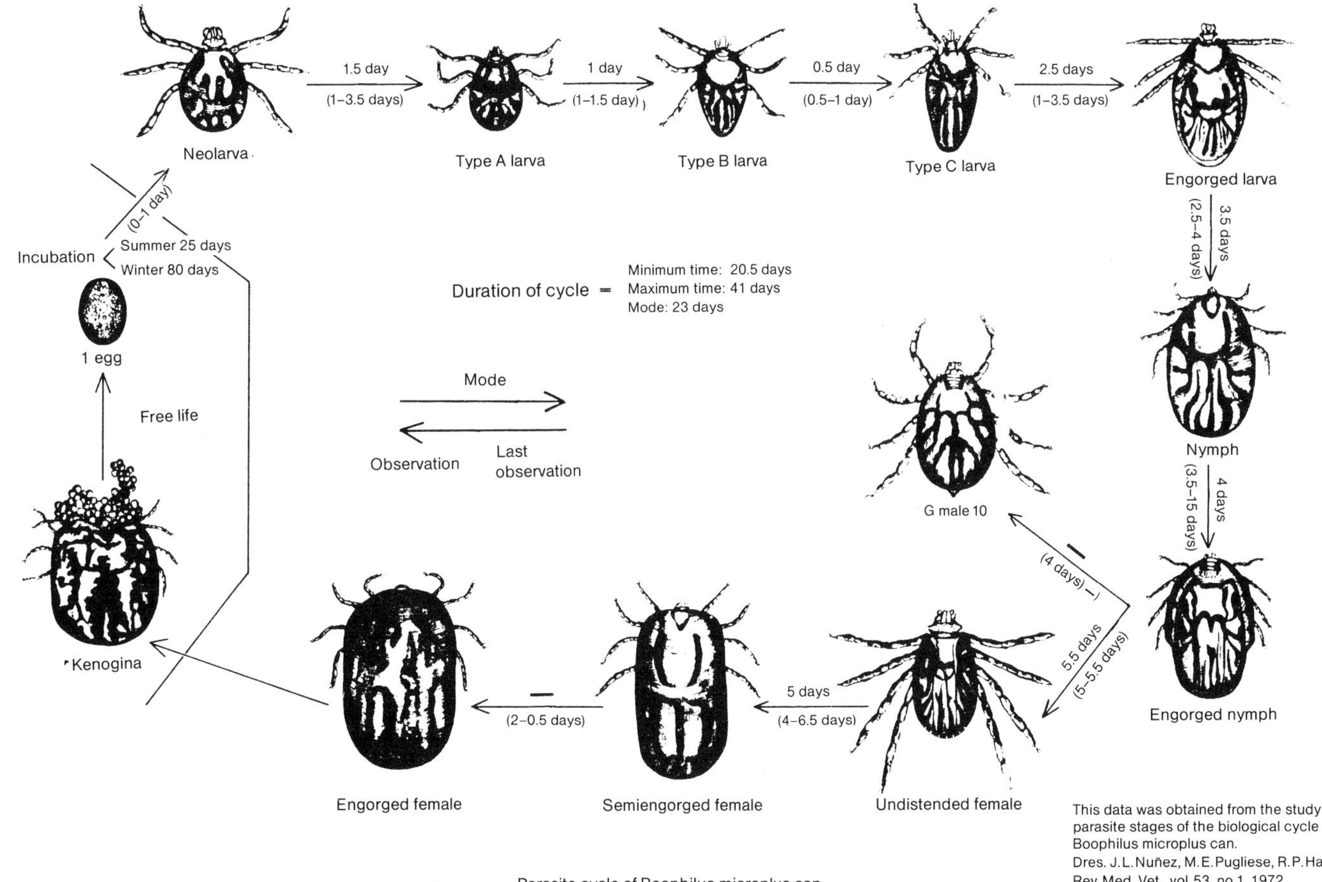

Fig. 4.6. Parasite cycle of *Boophilus microplus can* (Nuñez et al. 1972)

The first undistended female usually emerge 14.5 days after infestation and generally they do not wander, but attach themselves again to the host near the site where the engorged nymph from which they came was located.

Semi-Engorged Female

This name is given to the female after she starts feeding and growing until she is fertilized and finally when she detaches as the engorged female.

The undistended female, as has already been explained, attaches herself very near to the previous site of the engorged nymph and, now known as the semi-engorged female, starts growing: the growth is very slow at first, which is why it is very difficult to differentiate between the undistended female and the semi-engorged female for 1 or 2 days.

On the 3rd or 4th day, the weight increases by 80% compared to the undistended female; from then onwards, the development is very fast, increasing by 400% in the vast majority of cases between the 4th and the 5th day. At this stage of development the longitudinal grooves, two on the anterior half of the body and three on the posterior, are clearly seen dorsally. These grooves correspond to the insertion of the dorsoventral muscles.

Her colour is dark brown at the beginning, but somewhat lighter than in the case of the males. Later, when the parasite is 3 to 4 mm long, it becomes opaque.

Semi-engorged females can be clearly observed 17 − 18 days after infestation. We consider it important to note that in our experience, in all infestations some females are left unfertilized (we have noticed this phenomenon also while working on cellulae), and in such cases development ceases, the parasite remaining until it dies on its host. In no case have we observed parthenogenetic phenomena, that is, females fertilized without having mated with a male, although we sometimes observed small populations composed exclusively of semi-engorged females.

Engorged Female

This is the name given to the fully engorged female, who concludes her development and then detaches so as to lay her eggs off the host. In general terms, we have observed that the engorged female increases in size from 4 to 6 mm and waits for copulation, after which she reaches complete engorgement within a very short time, sometimes within hours, a fact which is also described by Legg (1930). The engorged female is ovoid in shape, greyish in colour and measures from 7 to 13 mm long × 4 to 8 mm wide. Her average weight is approximately 240 mg (225.38 − 266.33) according to Londt and Arthur (1975) and 274 mg according to Lahille (1917). The latter records that the average volume of water which the female displaces is 0.229 cm^3. Legg gives an average weight of 0.11 g.

In our study of the parasitic cycle of *Boophilus microplus* (Nuñez et al. 1972) we observed 3746 engorged females: the first one detached 20.5 days after infestation and the last after 41 days. The largest number detached on day 23: 535 specimens (14% out of the whole). An interesting detail (see Fig. 4.7) in the de-

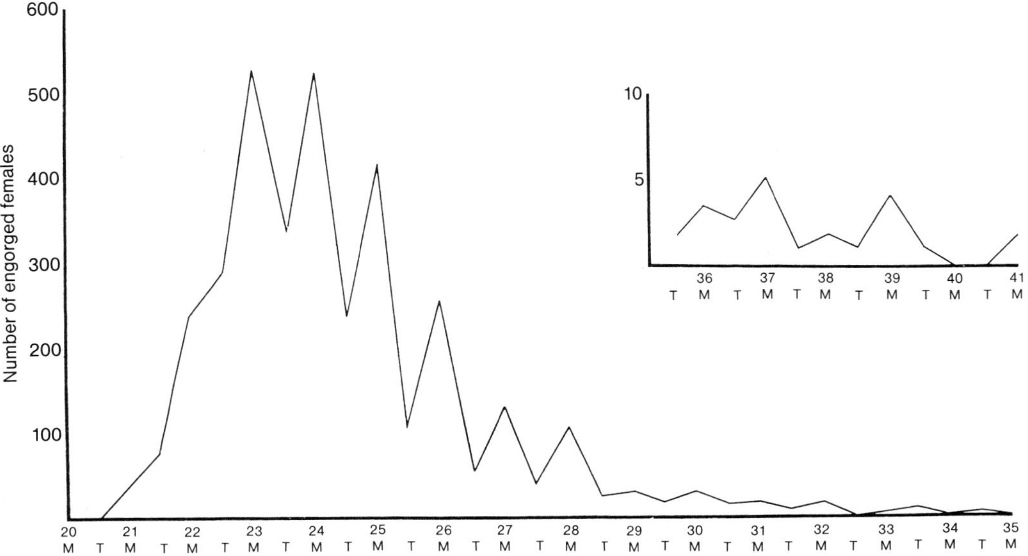

Fig. 4.7. Knock-down of engorged females. *M* morning; *T* afternoon. (Nuñez et al. 1972)

Fig. 4.8. *Boophilus microplus can.* Group of engorged females with some morphological alterations

tachment of engorged females is the remarkable tendency of the vast majority to do so in the morning rather than in the afternoon. Hitchcock (1954), made this very same observation. We have also observed a low percentage of ovigerous females bearing several morphological alterations (see Fig. 4.8).

Some of these engorged females laid eggs. The respective ovipositions resulted in a normal percentage of eclosion.

Some Thoughts on the Parasitic Cycle

On analyzing the graphs of both larval and nymphal stages and the whole life cycle (see Figs. 4.6 and 4.9 and Table 4.2), it is easy to see that each stage forms a typical frequency curve (Poisson n = 5) whose ends go further away from the mode as the cycle develops; the posterior end tends to do so more than the anterior. This phenomenon is clearly seen when certain stages, such as engorged nymphs, are studied.

Another subject which deserves greater attention is that in some stages, when attached to the anterior border of the ears, the tick develops very slowly (which may also be influenced by which part of the host's body the larva inhabits). After a detailed analysis of the available data, the idea that a plan for eradication by means of periodic treatments can be based upon data within the normal range is confirmed; but when the survival of only a few specimens is being judged, the distribution of these observed when studying each stage must also be taken into account.

This is of primary importance when evaluating the therapeutic action of an ixodicide, where scientific criteria must be combined with a clinical check-up; the conclusions must be a balance of all the factors involved, must avoid wrong interpretations, and combine realism and application.

Still another detail which must be taken into account is the small influence exercised by environmental factors upon the parasitic cycle.

The data given by Legg (1930) in Australia and by Sapre (1940) in India confirm that the differences are practically insignificant.

We also observed in our own work that the age of the animal, its physical condition, the severity of the infestation and the presence or absence of previous infestations had no influence upon the development of the life cycle.

Another characteristic which can be considered as constant is the high mortality rate observed throughout the parasitic cycle, ranging between 2 and 7% in the larval stage and from 20 and 40% in the adult.

In our work on this subject (Nuñez et al. 1972), in animals artificially infested with approximately 20,000 larvae, 3746 engorged females were collected: assuming that there was an equal number of males, the mortality rate would have been about 60%. These data partially confirm those which Legg (1930) and Hitchcock (1954) have reported.

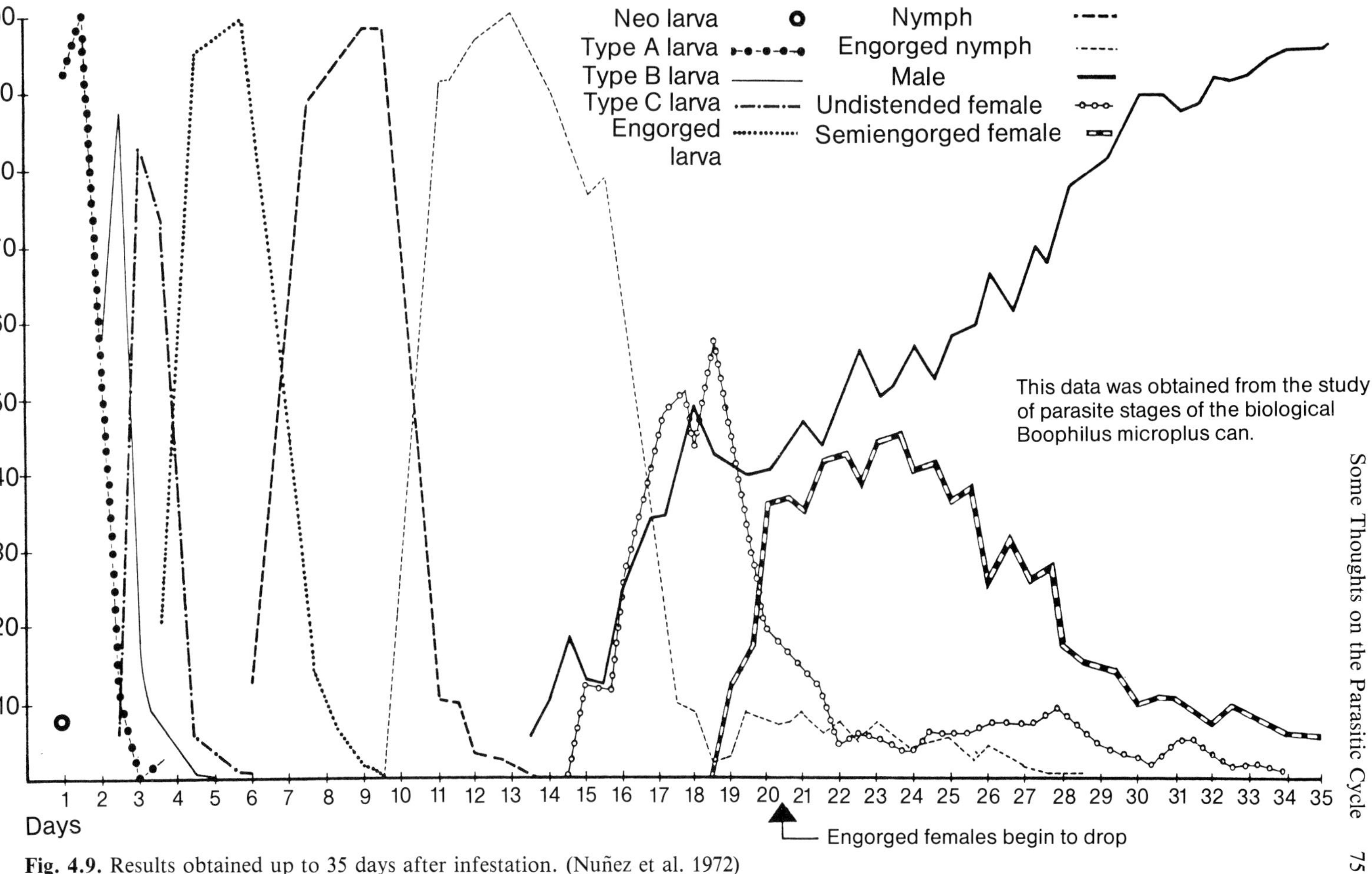

Fig. 4.9. Results obtained up to 35 days after infestation. (Nuñez et al. 1972)

Table 4.2. Parasite cycle of *Boophilus microplus can.* (in days)

Stage	First observation	Last observation	Mode	Mean	Variation	Standard deviation
Neolarva		1				
Larva type A	1	3.5	1.6	1.7	0.0038	0.195
Larva type B	2	5	2.5	2.5	0.05	0.215
Larva type C	2.5	6	3	3.4	0.08	0.278
Engorged larva	3.5	9.5	5.5	5.4	0.1	0.317
Nymph	6	13.6	9	8.7	0.14	0.375
Engorged nymph	9.5	23.5	13	13.9	1.75	0.322
Male	13.5					
Undistended female	14.5	34	18.5	19.7	4.2	2.051
Semi-engorged female	18.5	40.5	23.5	24.9	1.9	1.366
Engorged female	20.5	41	23	24.6	0.8	0.912

Table 4.3. Comparison of data obtained from different authors (Nuñez et al. 1972)

Authors	Moulting larvae		Moulting nymphs		Engorged females drop off		
	First	Last	First	Last	First	Last	Mode
Hooker et al. (1912) U.S.A.	6	9	14	19	22	29	22 – 25
Lahille (1917) Argentina	5	9	17	23	23	42	–
Legg (1930) summer, Australia	6	10	13	20	20	34	22
Legg (1930) winter, Australia	6	10	13	21	20	35	24
Sapre (1940) India	6	9	13	18	24	28	26
Tate (1941) Puerto Rico	9	11	14	19	20	27	22 – 23
Hitchcock (1954) Australia	5	9	13	21	19	36	23
Nuñez et al. (1972) Argentina	6	9.5	13.5	28.5	20.5	41	23

According to Nuñez (1972)

References

Alvarado RU, Gonzalez JC (1979) A postura e a viabilidada do *Boophilus microplus* (Canestrini, 1887) (Acarina, Ixodidae) en condiciones de laboratorio. Rev Latinoam Microbiol 21:31 – 36

Ault CN (1948) Investigaciones sobre las dificultades de combatir la garrapata *Boophilus microplus*. Rev Med Vet Buenos Aires 30:1 – 84

Boero JJ, d'Angelo (1946) Biología del *Boophilus microplus* garrapata común de los bovinos. Bol Tec Dir Gral Ganadería no 14

Davey RB, Garza J Jr, Thompson GD, Drummond RO (1980) Ovipositional biology of the southern cattle tick. *Boophilus microplus* (Acarina, Ixodidae) in the laboratory. J Med Entomol 17:117–121

Hitchcock LF (1954) Studies on the parasitic stages of the cattle tick *Boophilus microplus*. Can Aust J Zool 3:145–155

Hooker WA, Bishop C, Wood HP (1912) The life history and bionomics of some North American ticks. Bull US Dep Agric no 106

Ivancovich JC (1975) Bioecología de la garrapata del ganado *Boophilus microplus* (Canestrini, 1888). Rev Invest Agropecur INTA Ser IV 12:no 1, 1–54

Joan T (1940) La garrapata común del gonado vacuno. Boletín Fomento Ganadero, Año V, no 16

Lahille F (1917) Atlas de la garrapata transmisora de la tristeza. Bol Minist Agric, Tomo XXII no 2, p 243

Legg J (1930) Some observations on the life history of the cattle tick (*Boophilus australis*). Proc R Soc Queensl 41:121–132

Londt JGH, Arthur DR (1975) The structure and parasitic life cycle of *Boophilus microplus* (Canestrini, 1888) in South Africa (Acarina, Ixodidae). J Entomol Soc South Afr 38:321–340

Nuñez JL, Pugliese ME, Hayes RP (1972) *Boophilus microplus Can.* Estudios sobre los estadios parasitarios del ciclo biológico. Rev Med Vet Buenos Aires 53:no 1, 19–35

Quevedo JM (h), Gutierrez RO, Elizondo MJ (1960) Garrapaticidas, garrapatas y una técnica para su estudio. 2nd Congr Nac Vet, Buenos Aires, pp 107–118

Sapre SN (1940) The life history of *Boophilus australis* (Fuller). Indian J Vet Sci 10:346–353

Tate HD (1941) Biology of the tropical cattle tick and species of ticks in Puerto Rico with notes on the effects on ticks or arsenical dips. J Agric Univ Puerto Rico 75:1–24

V Therapeutics and Control

Principles of the Control of Ticks

The most efficient method to obtain control over *Boophilus microplus* consists in preventing the parasitic forms on the host from reaching the stage of engorged female, thus avoiding their dropping to the ground, oviposition and the consequent hatching of larvae that produce renewed infestations.

According to the life cycle already described, we see that under favourable conditions it is completed within an average of 22 – 23 days (Kearnan 1974) (Nuñez 1975).

Therefore the best way of interrupting the cycle is by using good ixodicide dips (Jones-Davies 1972) (Purchase 1955) (Rafyl and Maghalai 1959).

Theoretically, after the first dip, 99% of the ticks are destroyed, and thus another 3 weeks would have to go by before the engorged nymphs can develop on treated animals (Norris 1967).

However, under normal conditions and due to the protective residual effect of a good active ingredient, the larvae reinfestation will be delayed after the first dip for about 4 days. This means that the period of time between the dip and the reappearance of engorged females will be not less than 25 days.

Therefore, if we treat the animals a second time 3 weeks after the first dip, we would theoretically be anticipating the appearance of reproductive stages, obtaining a satisfactory control, very adequate for routine treatments.

Nevertheless, this method is too theoretical to obtain a clean herd intended for a parasite-free area (Newton 1967). It does not take into account that in the field, larvae infestation is continuous (Vercherkin et al. 1975) (Norris 1967) and that on each animal there are elements of different stages, many of them mutating during the first dip, and being protected by the cuticle of the previous stage, such as engorged larvae and engorged nymphs, that are more resistant to this treatment.

Thus the need arises for another method that will prevent this from occurring, for example, by repeating the second treatment 9 days after the first, or following the schedule of Fig. 5.1.

In conclusion, when trying to eliminate all stages completely, the intervals between dips must be shortened. The schedule shows that the desired effect is

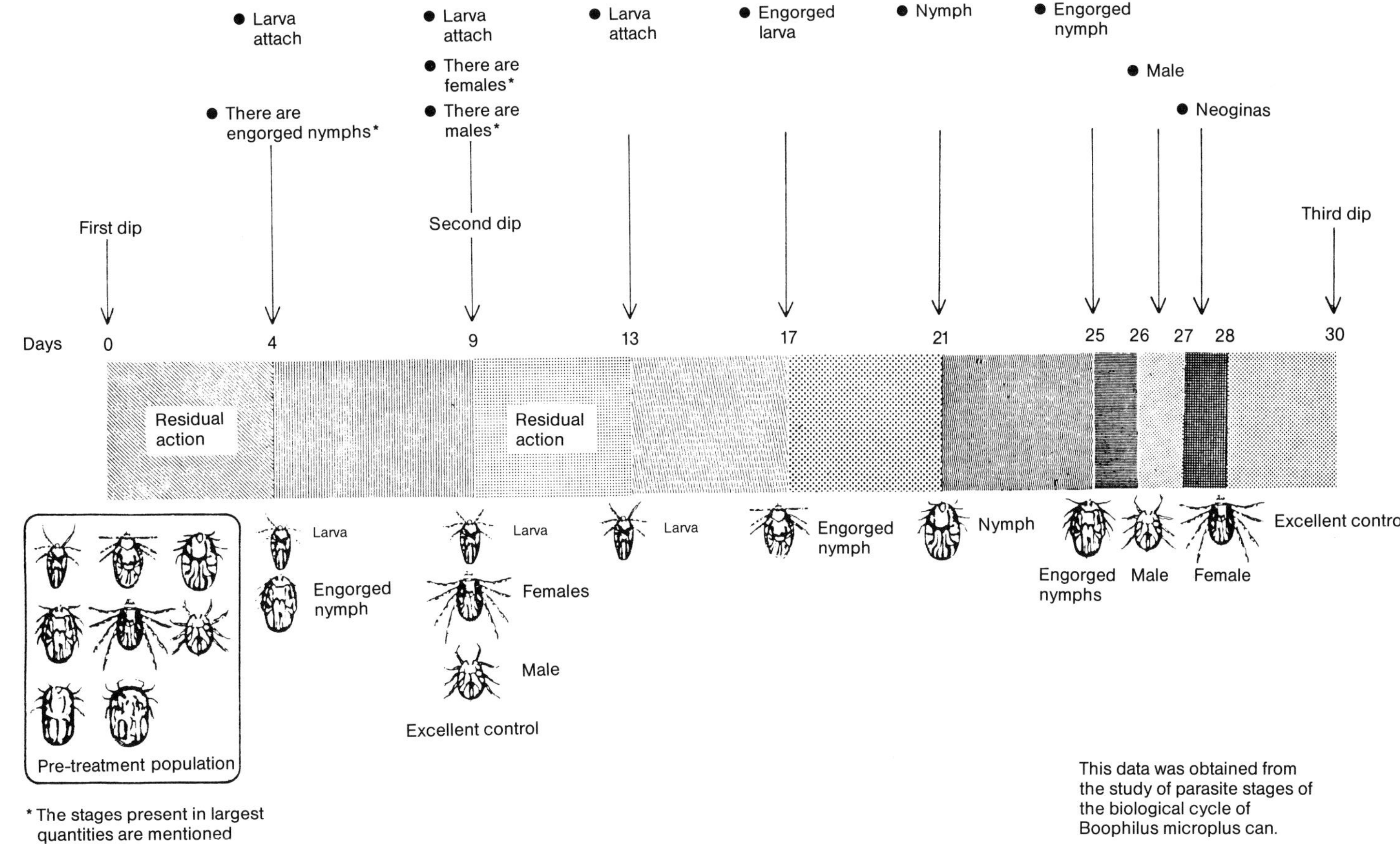

Fig. 5.1. Tick control by dipping

achieved by carrying out 3 dips in 21 days. If any stage is not affected in the first dip, it will be in the second or third. Wharton (1967) compared different intervals between dips, 21 to 28 days from the first treatment, using animals with similar degrees of infestation and the same organophosphorus ixodicide. He observed that in the first period only 622 engorged females dropped, while on application of the second dip with an interval of over 28 days, this drop increased considerably, up to 7800 engorged females.

The properties of the ixodicides used must be known, as not all of them have the same protective residual action.

Use of Chemical Products in the Control of Ticks

In countries such as Argentina, where ticks are an exotic species (Wells 1975), the only means of control are those practised by man, for ticks have practically no natural enemies, predatory or pathogenic, that can interpose a biological barrier (Bhatia 1974) to their rapid expansion.

Therefore, as already mentioned, it is essential to use chemical products toxic to ticks, called ixodicides (Newton 1967).

The most commonly used are concentrated liquids which, as well as containing an active ingredient or ixodicide, also includes in their formulation emulsifiers and solvents or moisturizing agents that play an important role in the quality of the product. The products generally come ready to be diluted in dipping tanks, in the proportions indicated by the manufacturer.

In order to proceed correctly it is necessary to pay special attention to the instructions on the label, which also include safety precautions for the operator, thus preventing risks during their use.

Properties of a Good Ixodicide

A good ixodicide should have four basic properties.

1. Direct effective action not less than 99.9%.
2. Residual effect.
3. Action on oviposition.
4. Stability.

In our experience (Nuñez 1975), the first two characteristics are the most difficult to evaluate, as they depend upon a thorough clinical examination of the treated animals, where the presence of only one element must be studied carefully, using frequency curves for each stage of the life cycle.

The third characteristic, relatively easy to evaluate, is based on laboratory trials where the treated engorged females are kept in special stoves at adequate temperatures over a sufficient length of time for them to complete oviposition, it being determined at the end if this was altered or not.

Stability is the condition by which an ixodicide keeps active in a dip, even after a large number of animals have gone through it. The reason for this being that, during their passage, different organic substances, such as earth, cow dung and hair, are incorporated into the dip and might alter it. If the product is stable, its therapeutic concentration should not alter.

For this specification the product must prove its persistence during not less than 180 days, during this time, of course, following instructions for reinforcement or replenishment as indicated by the manufacturers.

Exhaustion of Ixodicides in Dips

The phenomenon known by the generic name of exhaustion or stripping consists of the chemical changes in the active ingredient when it becomes inactive. This results in a loss of biological efficacy and is shown in therapeutic failures in the fight for tick control.

Thus the quantity of compound in the dip decreases, as well as the amount that each animal can carry.

This depletion may occur after the passage of the animals through the dip, for each one carries the active drug on its body surface: if the product is an emulsifier, in the form of droplets, if it is a suspension, as small particles.

The animals are left to dry on draining boards placed immediately after the dip, so that excess liquid can run back into the dipping tank. In spite of this, the liquid becomes weakened, and the ingredient less active, thus causing a weakening of the concentration of ixodicide.

Although so far the mechanism of selective absorption or exhaustion in dips has not been fully explained, it is known that it can increase with the insolubility of the components in water. Other factors might increase the rate of exhaustion: water temperature, length of hair, volume of the tank in relation to the number of animals dipped, degree of contamination (Newton 1967).

This partly explains the need for reinforcements at concentrations sometimes 200% above the original.

Another form of stripping is caused by the physical changes in the size of the particles, which tend to decant on the bottom. The decanting of dip suspensions is quite normal, although it is not so marked when the dispersing agents of the formulation are of good quality and the milling of the particles is correct (Elder et al. 1980).

The contaminating particles stirred up by the animals can absorb the chemical particles on the surface, speeding up the decantation and thus weakening the concentration of the active ingredient.

In this respect we mention the work done with DDT and BHC by Dingle (1956). He established important differences in the speed of sedimentation between recently prepared suspensions and others to which he added the contents of a dirty tank, thus producing an increase in decantation. He also noted that these levels of sedimentation could increase if the concentration of the active ingredient had decreased to the minimum.

Although emulsions have a much smaller degree of decantation due to the droplets having equal electric surface charges and thus repelling each other and maintaining a continuous movement, they may be affected by the salinity of the water, particularly by the presence of sodium chloride.

If rupture of the emulsion occurs, the results would be a greater concentration of ixodicide floating on the surface, which, besides increasing the risk of intoxication, is subject to greater dragging during the passage of the first animals.

In order to check these effects the formulations must include good miscible oils. These employ synthetic emulsifiers that facilitate a satisfactory mixture with waters containing a high degree of sodium chloride or that are brackish for other reasons.

Lastly, the wastage in tanks caused by the action of bacteria such as *Pseudomonas* spp. must be considered.

From early times these bacteria have been known to cause stripping of arsenical solutions by oxidization of sodium arsenite to sodium arseniate, a compound without biological activity. These bacteria are also responsible for the loss of biological efficacy of BHC, as can be seen in the corresponding chapter, due to hydrogen not being liberated during the development of anaerobic bacteria which effect dechlorination.

From all this we see quite clearly the imperative necessity of carrying out regular analysis of dip samples; in order to know the exact concentration, and thus avoid the use of dips with a low degree of active ingredient that might facilitate the selection of tolerant strains.

Methods of Applying Ixodicides

As Barnett (1969) points out, the application of tickicides to the host has important advantages. For one, it protects the animals from the traumatic effect of a sting and the toxic effect of the transmission of anaplasmosis and babesiosis, instantly destroying all the ticks carried. For another, it cleans the infested field, for it permits the host to become loaded with the ticks present in the field and then destroys them by means of the dip. In due course this method makes it possible to free the pastures of larvae.

Without doubt, the current methods of applying ixodicides are by immersion and spraying.

Fig. 5.2. Plunge dipping tank

Immersion Dips

The first of the methods known from early times has proved to be the most efficient. With this method, the complete wetting of the whole body surface is obtained, allowing intimate contact of the compound with all the stages present on the host. An additional advantage is the great speed of the operation, particularly in large round-ups (see Fig. 5.2.).

The material used for the construction of dips is very varied: stone, brickwork, hard wood, steel and, in some countries, fibre glass has been tried.

The best dips are made of brickwork and concrete, as they are less exposed to cracks and filtrations. Their location is important, preferably in the centre of the establishment in order to avoid moving the animals long distances.

It is also necessary to study the type of soil where they are to be built, to avoid cracking or the proximity of springs that might cause filtrations.

Normally an immersion dip is formed of three sections: (1) the approach section, (2) the tank, (3) the drainage section.

The first section consists of a funnel, generally the continuation of a conveniently conditioned race, in order to assure the fluid passage of animals. It is advisable to build in this section a foot-wash with a rough floor surface, so that the hooves open and rid the animals of dirt. It is best to try and keep the tickicide liquid as clean as possible, for which it is good to pave the access to the dipping tank.

The tank itself must be long enough to allow the animals to swim for not less than 15 s, and deep enough for them not to touch the bottom thus not submerging their heads completely; generally it should be not less than 2 m deep. As well as assuring the complete immersion of their heads without suffering any damage, the tank should have a capacity varying between 13,000 and 15,000 l.

The width of the surface should be wide enough to dip horned animals, and the floor must be not less than 45 cm wide to permit a correct stirring of the liquid and to avoid deposition of the active ingredient.

To facilitate exit, the tank must have a ramp with an angle of 20° and steps every half metre.

The drainage area is a corral placed after the tank, and normally divided into two sections with a capacity for 50 – 60 animals. A paved floor ensures the return to the tank of the surplus liquid carried out by the treated animals.

The dipping tank should be completely roofed, to prevent rain water from dropping into the dip, and to reduce evaporation, thus preventing dilution of the ixodicide (see Fig. 5.3). Roofs are most commonly made of thatch or metal sheets. The tank should also be provided with two sumps for decanting the returning liquid and retaining coarser particles such as earth and dung. These sumps should be situated between the drainage area and the dipping vat.

An auxiliary tank of 1000 – 2000 l capacity next to the dipping tank greatly helps the replenishing of water and even, in many cases, can be used to premix the product before reinforcement.

Lastly we mention that there must be a good water supply near by, avoiding water pumps that might hamper the free flow. Normally it is convenient to have the dips near a large source of stored water (e.g., an Australian tank).

Dipping the Animals

The best time to carry out the dipping is early in the morning, in dry and cool weather, avoiding rainy days.

Prolonged round-ups should be avoided, and the cattle kept calm. If this is not possible, it is preferable to round them up the previous afternoon, thus avoiding dipping tired or thirsty animals, which could lead to their drinking the liquid. For this reason it is advisable to have drinking troughs in the waiting corrals.

One of the ways of speeding the passage of the animals through the dip is by taking them in groups of equal age and sex.

It is a good idea to keep a record of the number of animals treated and the product used, as well as regular samplings, in order to be certain of the correct concentration of the ixodicide present in the dipping tank.

Fig. 5.3 A, B. Roofed dipping tanks

Handling of Immersion Dips

Every time a wash is changed, the tank should be thoroughly cleaned, removing all the dung and mud that is normally deposited on the bottom, and at the same time checking the walls for filtration and carrying out necessary repairs.

The next step is to measure the volume of the tank for which a 200-l drum can be used. At the same time, a graduated rule must be made for measuring the contents of the tank.

After filling the dipping tank with water, the product is added in the exact proportions for preparing the dip wash as indicated on the label of the container, while observing the rate of replenishment and reinforcement. The contents of the tank must be stirred vigorously. This is a very important point. It should be done by hand before dipping and complemented by the passage of some 50 animals, that should be dipped again once the dip wash has been thoroughly mixed.

The time each animal remains in the dip must be strictly controlled and never be less than 15 s. Their heads must be submerged two or three times, taking care that the inside of the ears is completely wetted.

Once the operation is finished, the level of liquid should be accurately measured in order to determine its increase or decrease and correct it for the next treatment.

Disadvantage of Plunge Dips

Although there is no doubt that this method is the most efficient, it has some disadvantages, such as the need for larger quantities of water and above all, the difficulties in maintaining the correct concentration of the ixodicide (Elder et al. 1980).

Also, the need to empty the dipping tank to clean it, or to change the compound.

Finally, due to their greater predisposition to traumatisms, this system is not the best for pregnant females, nor for milk cows, due to their pendulous udders.

Procedure for Correct Sampling

The liquid must be stirred thoroughly by hand for 10 min, and as said above, by means of the passage of not less than 50 animals in order to obtain the correct homogenization of the contents. The sample should be taken from a medium depth and immediately after the passage of the last animal.

Spraying

This widely used method has been applied in many countries due to the direct advantages it offers. Not only does it permit an exact knowledge of the concentration of the active ingredient, but it also avoids the risk of intoxication due to error, and the use of recently prepared dips.

The method is also economical as it is only necessary to prepare the quantity required according to the number of animals to be treated. It also allows the application of different types of tickicides according to the species of ticks present in different seasons, specially important in zones where several species exist.

Another important factor is the lower cost of operation, since it does not require the expensive installations of dipping tanks. The system is also more practical because the installations are usually portable. When used for cattle, it causes less traumatisms than plunge dips.

Despite the advantages listed, with this system the wetting of the animals is not as complete, and the time of contact with the ixodicide is shorter.

Added to the advantages listed, mechanical aspersion systems designed allow a speedy passage of animals at a rate of 700 per hour (see Figs. 5.4 and 5.5).

Spraying consists essentially of the passage of cattle through a spraying race specially equipped with a piping system and nozzles placed at regular intervals through which the liquid comes out in the form of a fine spray. The wash from tanks of 1500 to 2000 l capacity is pumped up and pressure-fed through tubes and nozzles on the frames and horizontal pipes of the spraying race, placed in such a way that the animals get an overall wetting when passing through. The system also has an internal crush race solidly built to ensure that the animals keep a certain distance from the nozzles and thus avoid damaging them. This is an essential feature which ensures that the cattle pass through the race in single file at the correct distance from the nozzles and thus receive even and complete coverage by the spray. The crush race consists of four pairs of uprights embedded into a concrete platform so that the distance between the centres of each pair of uprights is 46 cm at floor level and 91 cm at the top. The external walls of the

Fig. 5.4. Spray race

Fig. 5.5. Spraying a cow

system are built of brick or concrete in order to prevent the spray from being blown away by the wind.

The entrance race is also made of concrete with a rugged surface. It should be provided with a foot bath to ensure that the large particles of mud are loosened, thus preventing them from blocking the nozzles during recirculation of the wash. The floor at the other end of the spraying race should be sloped, acting as drainage area and facilitating the recovery of wash and its return to the main tank. For this reason it is advisable that this path be approximately 20 to 30 m long.

Although the specific details for the construction of spraying races are found in technical manuals, we give below some practical details to be taken into account prior to the installation of the system.

The first important point is the location. It must be built in a direction opposite to that of the prevailing winds, thus avoiding loss of wash.

Another important detail is that the spraying race should be at least 100 m from dairy buildings, thus avoiding all possibility of contaminating the milk. It is advisable to build it near a source of abundant water and in an area free of trees and bushes, to avoid leaves falling into the tank.

For the cattle to grow used to the system, it is advisable to introduce it gradually, making them pass through the race several times without spraying and preferably then once again with the water spray at low pressure.

The motor for the pump should be between 8 and 10 HP. A 7 HP electric motor can also be used or any motor that can supply power, for example a tractor or jeep.

Hand Spraying

This simple system is based on the application of the ixodicide by means of a hand spraying pump. It can also be used as a complement to mechanical spraying or plunge dips. It is generally used in establishments with a small number of animals.

The success of the treatment depends on the methodical application over all the body surface, paying particular attention to those areas of difficult access such as the insides of ears, base of the tail, belly, etc. For this reason it is important to have a suitable corral or race, and to secure the animal during treatment.

Generally the races used to immobilize cattle are made of wood which might impede the access to the nozzles. If this system is used it is advisable to use narrow planks for building these races.

Sliding doors can be used to separate the animals. As each one enters the race, the doors are put in place, preventing the cattle from hiding their heads between the legs of the one in front, which would make the spraying of their heads and hind quarters difficult and the operation slow and inefficient.

A more economical alternative to these doors could be poles placed between each animal as it enters the race.

Another system which has given excellent results is the construction of two races placed in a V shape, each one capable of holding ten animals, using them alternately. The cattle move into them from a small crush. The pump and tank containing the wash are built between the two races. One man handles the pump and two others the two nozzles on either side of the race. Meanwhile, another two men fill the second race with cattle. Including the time needed to prepare the wash, by this method 60 animals can easily be sprayed in 1 h. This double race system is practical when over 200 animals are to be sprayed. If the herd is smaller, then one race alone is quite sufficient.

In establishments where there are not more than ten animals and they are used to being handled, a simple three-post crush is satisfactory. In the three-post crush the beast is haltered and led into the "V" of the crush. The halter rope is then tied to the apex post. This crush allows free movement all round the animal and maximum freedom from obstructions when spraying.

Another point to be considered is the use of the exact proportions of concentrate and water. The product must be mixed in the tank with a small amount of water, adding the remainder as indicated in the instructions. This mixture must be stirred for a few minutes and repeated again at regular intervals in order to obtain a correct homogenization of the active ingredient in the water.

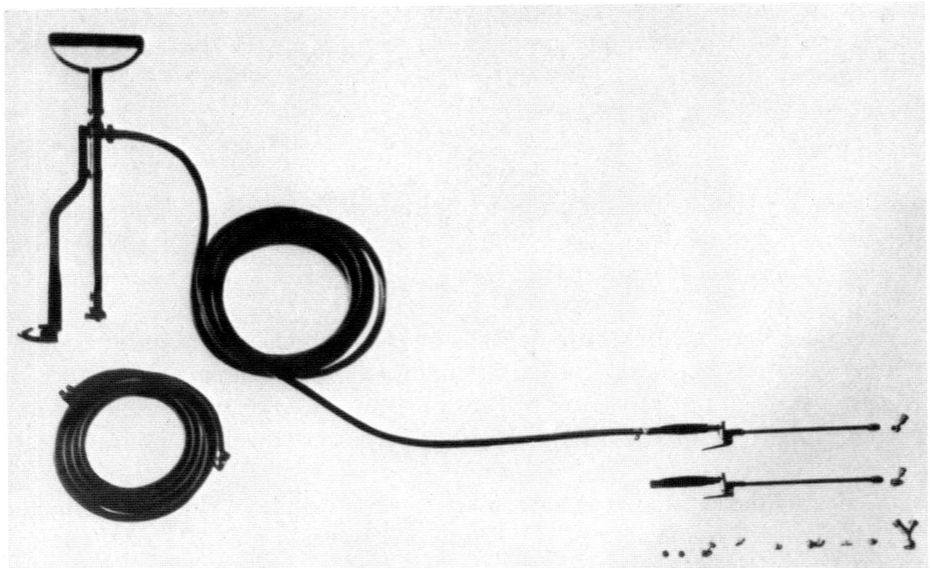

Fig. 5.6. Hand pump

Fig. 5.7. Hand pump

The only advantage this last method offers over the two previous ones is that there is no possibility of the wash returning to the tank, thus eliminating the stripping of the active ingredient.

Finally, the last requirement is the use of an efficient hand pump. It must provide a continuous pressure of not less than 7 kg cm^{-2}. The nozzle must let out an adequate amount of wash, between 2 and 4 l min^{-1}. The use of nozzles that only let out a light mist must be avoided. The nozzle pipe should be strong enough not to break, and long enough to reach all parts of the animal without difficulty and be comfortable for the operator.

The hose must also be long enough to avoid having to move the tank (see Figs. 5.6 and 5.7, hand pumps).

Practical Tips for Application

As previously stated, in order to obtain complete wetting of the entire body surface of the animals, using the minimal amount of wash, the application of an ixodicide must be done carefully and methodically.

The operator must keep the spraying nozzle 15 to 30 cm away from the animal. First the back, from neck to base of the tail, must be sprayed, followed by flanks and both sides of the torax. If it is applied in zig-zag, the hairs can retain even more liquid. Then continue with the internal and external surfaces of the legs, chest, lower belly, udder or scrotum. The tail and rump must be sprayed very carefully, as many ticks dwell in these areas. Finally, neck and head, face and orbits are sprayed, leaving the interior of the ears for the end.

In Argentina spraying against ticks is not permitted in sanitary campaign areas. Its use is restricted to treatment for the control of cattle mange, lice and other insects. Nevertheless, in other countries, such as Africa for example, it has been particularly promoted for treating anatomical parts of difficult access, that can be more effectively sprayed, such as interior of ears, base of horns and perineum, where species other than *Boophilus microplus* tend to seek refuge.

Description of Pumps

A CS1 spraying pump is adequate for herds of up to 20 animals; a CS2 for herds of up to 60 heads; a Spray Well pump by Cooper is suitable for large herds.

This pump consists of a double-action piston, its base and body being made of cast iron. It supplies 2.2 to 3.4 l min^{-1} with a pressure of 7 kg cm^{-2}. The valves are of synthetic rubber and the piston has a cylinder of brass.

Although this pump works with two levers, one man can easily handle it. Its base is perforated so it can be set up on a wooden plank.

Disadvantages of the System

As Do Couto Souza (1970) points out, the motor is subject to breaking down. The particles of toxic liquid that escape might touch those people working with the wash.

The piping system must be kept clean and this requires considerable maintenance. Perhaps the most serious problem is the impossibility of obtaining a complete wetting, as the animals instinctively lower their ears and tails when going through the curtain of liquid, thus preventing direct contact of the wash with areas such as ears, tail, escutcheon and axillae, which are subject to parasite infestation.

Application of Systemic Ixodicides

This is based on the diffusion of certain active ingredients through the tissues and skin of the animals treated with products administered orally or by injection, so that the parasitic stages feeding on their host are destroyed by intoxication by the drug ingested.

This operation, practical when compared with the difficulties of spraying and plunge dips, has been the subject of much research.

The first experiments were carried out during the 1940's by treating rabbits orally against ectoparasites (Lindquist 1944, Demeillon 1946).

In spite of the good results obtained by Rayfi et al. (1956) with Dieldrin applied orally at $50\,\text{mg kg}^{-1}$, its persistence makes it undesirable, as man ingests dangerous residues when eating meat from treated animals, and also because of the possibility of development or resistant strains. The first attempts at tick control were by Wilson in 1948, using different concentrations of BHC and DDT, and obtaining good results in the treatment of *R. appendiculatus*.

In 1956 Roulston, by injecting cattle subcutaneously with different compounds and then exposing them to natural infestation of *Boophilus microplus*, demonstrated that only one injection of $25\,\text{mg kg}^{-1}$ of Aldrin, Dieldrin or Lindane was effective against ticks. Although some were able to complete their life cycle, the residual effect lasted for over 2 months.

Two other active ingredients can be mentioned: Avermectin and Closantel. Both of them have proved to have a systemic therapeutic effect; however, to obtain significant levels of efficacy against bovine ectoparasites, repeated high doses are required.

References

Barnett ST (1961) The control of ticks on livestock. FAO Agric Stud 54:1 – 115

Bhatia HM (1974) Ixodicides in animal tick control. Vet Bull 44:5:273 – 279

Camoens JK (1977) The control of tick infestation on intensive dairy farm. Malays Vet J 6:111 – 124

Demeillon B (1946) Effect on some blood-sucking arthropods of gammexane when fed to a rabbit. Nature (London) 158:839 – 840

Dingle J (1956) Settling rates of DDT and BHC suspensions in cattle dipping baths. Aust Vet J 32:132 – 137

Do Couto Souza Octavio (1970) Combate aos parasitos externos do gado. Gaz Agric 22:252:137 – 144

Elder JK, Emmerson FR, Kearnan JF, Waters KS, Dunwell GH, Morris RS, Knott SG (1980) A survey concerning cattle tick control in Queensland. Aust Vet J 56:212 – 217

Kearnan JF (1974) Cattle tick control and a history of tick resistance. Queensl Agric J 100 (5):177 – 182

Jones-Davies WJ (1972) Tick control and a history of tick resistance. Rhod Vet J 2 (61):53 – 59

Lindquist AW (1944) Mortality of bedbugs on rabbits, given oral doses of DDT and pyrethrum. J Econ Entomol 46:610 – 614

Lombardo RA (1976) Importancia socio-económica del problema de la garrapata en las Américas. 8a. Reunión Interamericana sobre el control de la fiebre aftosa y otras zoonosis. Publicación Científica Org Panam Salud 316:86 – 97

Newton LG (1967) Acaricide resistance and cattle tick control. Aust Vet J 43:389 – 393

Norris KR (1967) Strategic dipping for the control of the cattle tick *B. microplus* (Can) in South Queensland. Aust J Agric Res 8:768 – 787

Nuñez JL (1975) La lucha contra la garrapata. Disertación durante las III. Jornadas Uruguayas de Buiatría, Paysandú

Purchase HS (1955) Some thoughts on ticks and their practical control. Bull Epizoot Dis Afr 3:266 – 330

Rafyl A, Maghalai G (1959) La lutte contre les argasides, gale psoroptique et phtiriase des moutons par l'administration des insecticides par la voie bucalle. Bull Soc Pathol Exot 52:39 – 42

Roulston WJ (1956) The effect of some chlorinated hydrocarbons as systemic acaricides against the cattle tick *B. microplus* (Can). Aust J Agric Res 7:608 – 624

Shaw RD (1969) Tick control on domestic animals. Trop Sci 2:113 – 119

Vecherkin SS, Puzil AD, Duisheev AD, Romakhov VG (1975) The outlook for eradication of piroplasmosis in the Jirgiz SST. Veterinariya (Moscow) 3:65 – 66

Wells EA (1975) Methods of tick control. Int Centre Trop Agric (CIAT), Cali, Colombia, pp 90 – 92

Wharton RH (1967) Acaricide resistance and cattle tick control. Aust Vet J 43:394 – 397

Wharton RH (1974) Tick with special emphasis on *Boophilus microplus*. In: Control of arthropods by Pal R, Wharton RH (eds) Plenum Publishing Corp, New York, pp 35 – 52

Wilson SG (1948) The effect on ticks and glossina of feeding cattle on BHC or DDT. Bull Entomol Res 39:423 – 434

Therapeutics

Evolution of Tickicide Drugs

Since ancient times man has observed the parasitic effect of certain natural products of vegetable and mineral origin. This paved the way for the first empiric therapeutics from which, in later years, the first scientific ixodicides developed.

Towards the end of the last century, such active ingredients as tobacco extracts, sulphur, oils and a few other compounds were already available. During the first decades of the 20th Century, arsenic and its salts were added to those already known. Later, in 1940, the ixodicide properties of organochloride compounds were discovered and were intensively used until they were restricted in the 1960's.

From them on, there were important developments in therapeutics. For example, the development of organophosphate compounds during the 1960's, by means of which important progress was made in the fight against ticks.

After 1970 new lines of therapeutics were developed and new compounds, such as formamidines and cycloamidines, were discovered.

Latterly, research into new drugs has allowed laboratories to work with new active ingredients such as pyrethroids, which have a longer residual effect and are less toxic to man and animals.

Although the ideal tickicide has not yet been synthesized, parasitic pharmacology has numerous by-products which, due to their reduced toxicity and therapeutic efficacy, provide sufficient means of making important progress in the control and erradication of ticks.

In this chapter we will study all the compounds used for this purpose, describing them and grouping them according to their chemical nature.

Arsenical Compounds

The first attempts made last century to combat ticks produced only poor results, using different substances empirically such as oils, paraffin, crude oil, tobacco solutions, sulphur compounds, lime, etc.

Effective chemical tick control commenced at the end of the last century with the application of arsenical solutions in plunge dips which had already been applied successfully for combating cattle and sheep mange (Australia Res. 1919).

In 1895 the use of arsenical solutions was recorded in Australia (Bekker 1960) and in 1902 in South Africa (Harrison et al. 1973a).

In 1906 the official American campaign against ticks was started, applying severe controls. Dip washes with arsenical solutions were applied fortnightly to cattle, horses, sheep and goats in an area of approximately 2 million km^2. The campaign ended successfully in 1943, but still continues as a quarantine measure when livestock is imported from Mexico.

Favoured by their low costs, solubility in water and the simple method of analysis for determining their concentration, arsenical solutions were so extensively used, that investigation of new active ingredients was delayed for more than 40 years.

The active substance of the first arsenical solutions used as tickicides was white arsenic or arsenic anhydride, now known as arsenic trioxide (AS_2O_3). To increase its solubility in water it was combined with sodium carbonate (CO_3Na_2), or with caustic soda (HONa), forming sodium metarsenic salt.

The preparation was so simple that the use of home-made formulations such as those described below, quickly became widespread:

White arsenic	2 kg
Sodium carbonate	1 kg
Water	50 l

or

White arsenic	2 kg
Caustic soda	8 kg
Sodium carbonate	10 kg

Each of these formulations gave 1000 l of dipwash at a concentration of 0.25% (Ministerio de Ganadería y Agricultura, 1941). Their use could be toxic for the operator as well as for the animals treated.

Other mixtures known were those of sodium arsenite and adjuvants such as oil, carbolic acids and soap. With time it was realized that arsenic was the principal agent and therefore the adjuvant substances were left out.

In this period the washes had concentrations of 0.16 to 0.20% of arsenic trioxide (AS_2O_3), obtaining an efficient control when the dips were carried out at weekly intervals (Metcalf 1955).

The action of arsenical compounds was studied in insects, but it is thought to be similar in ticks. Arsenical compounds are mainly absorbed orally and through the outer cuticle. They affect the mechanism of cell respiration by inhibiting the respiratory enzymes and causing the destruction of protoplasm (Ministerio de Agricultura de la Nación Argentina 1941).

On the other hand, Fink (1927) proved that the sulfhydrilic groups acted as recipients of the arsenic ion. Also, due to their high content of sulfhydrilic substances, amino acids such as glutathion and cisteine were protective, in turn, against the effect of arsenicals, which would explain – as will be seen in Chapter 8 on Resistance – the relation between resistant tick strains and the high content of these amino acids.

The sequence of symptoms on insects after ingestion of arsenic is similar in the different species. In general, the symptoms are: a gradual decrease of motility, loss of balance, asthenia, suppression of response to stimulation, and death (Munson and Yeager 1947).

The histopathological effects 3 h after ingesting the toxic compound are characterized by disintegration of the middle stratum of the intestinal epithelium with loss of the striated edge, dissolution of the cell membrane, cytoplasm, vacuolation and nucleus changes with dispersion and dissolution of chromatin (Negherbon 1959).

These effects were studied with lead arseniate. Similar lesions are produced by calcium arseniates and arsenic trioxide.

Regarding the toxicity of arsenicals, it must be remembered that there is almost no difference between the therapeutic and toxic dosis.

As shown in trials carried out by Reeves (1925) on sheep, cattle and horses, with oral application at low doses, arsenicals do not accumulate in the organism.

Nevertheless, the use of arsenical solutions in plunge dips can be dangerous due to the possible skin absorption, acute poisoning occurring within 1 or 2 days.

Lesser symptoms of intoxication are the presence of skin blisters and abrasions, depilated areas and digestive disorders such as diarrhea and vomiting. On damp days normal doses can be toxic, due to the longer period of time that the active ingredient remains on the skin.

Much attention must be given to the disposal of the dipwash due to the danger of contamination of rivers or the soil itself. For this reason arsenical compounds should not be used in spray dips. Residues on grass may be very highly toxic to the animals that eat it.

Brooks (1947) demonstrated that feeding cattle with alfalfa containing 650 ppm of the active ingredient proved fatal for them.

It has been known for some time that the protective residual period is very short: 2 days. In 1905 trials were carried out with *Boophilus decoloratus* larvae on the skin of cattle. They showed that 2 days after dipping with arsenic concentrations of 0.13 and 0.15%, the effect was lethal to all the larvae; 4 days later the effect decreased; and 7 days later it was null.

As from 1938 the first failures appeared in South Africa, and some years later experiments carried out in Australia proved the existence of resistant ticks (Hitchcock and Mackerras 1947).

Around 1940 similar difficulties were encountered in Argentina, and − as mentioned in Chapter 8 on Resistance − Ault (1948) carried out a series of studies to try and determine the causes of the problem. He arrived at revealing conclusions, the main points of which we transcribe below:

1. Undistended females and semi-engorged females are not easily affected by arsenical solutions;
2. no difference in efficacy was found when the dipping time was extended from 12 s to 2 min;

3. arsenical dips give better results when the wetting periods are prolonged, although the risk of intoxication increases;
4. a concentration of 0.24% is only slightly superior to one of 0.20%;
5. the addition of various substances (caustic potash 0.10%, sodium chloride 1.75%, sodium silicate 0.14%, kerosene 0.5% and nicotine sulphate 0.027%) did not increase the therapeutic efficacy of arsenical compounds;
6. increasing the frequency of dips with intervals of 4 to 5 days did not improve efficacy;
7. the minimal lethal dosis of arsenic for the engorged female was less than 0.05 mg.

Organochloride Compounds

In view of the repeated failure of arsenicals to control ticks, during World War II studies of the new organochloride compounds were made. These had shown excellent effects on various insects harmful to man.

The insecticidal properties of DDT which, like those of BHC, had remained unknown since the discovery of these compounds at the beginning of the century, were discovered in 1939. They proved to have a broad spectrum of action over a wide variety of insects, as well as a residual effect never seen in any other type of drug.

Thus, the organochloride compounds were rapidly incorporated into the reduced number of therapeutics available at the time, constituting a real revolution in the methods for controlling insects and arachnids.

DDT

DDT is derived from chlorobenzene, dichlordiphenyl-trichloroethane or, more precisely, 2.2-bis(p-chlorophenyl)-1,1,1-trichloroethane. Its molecular weight is 354.5.

Structural Formula:

Chemical Formula: $C_{14}H_9Cl_5$

It has an amorphous or solid, cream-coloured powder, insoluble in water and soluble in organic solvents. It is stable in a neutral medium, but in an alkaline medium it hydrolyses quickly, liberating HCl. Nevertheless, DDT is a stable compound under natural conditions such as light and air.

Synthesis:

$$Cl_3CCHO + Cl-\!\!\bigcirc \xrightarrow{H_2SO_4}$$

$$Cl-\!\!\bigcirc\!\!-CHOHCCl_3 + Cl-\!\!\bigcirc \xrightarrow{H_2SO_4} DDT$$

Formulations

DDT has the characteristic of being very versatile to formulate: wettable powders, emulsions and solutions in various solvents, aerosols and gases.

The most common formulations used in Argentina for the control of ticks were the wettable powders at a concentration of 50% (García Mata 1947).

This formulation was mixed with inert ingredients to facilitate milling and dispersion, with aluminium silicates to delay sedimentation in the dipping tank, and with crescylic acid to avoid degradation by its inhibitory effect on bacterial growth.

With the usual concentrations, good action was obtained against young tick forms on their host, but not against the more resistant engorged larvae and engorged nymphs, which continued their cycle. With the mature tick stages, the action of this formulation was irregular.

The residual effect lasted approximately 3 weeks, disappearing completely after 25 to 30 days. The period was reduced if the concentration of active ingredient was decreased by dipping a large number of animals.

Spray formulations were used in other countries: we can mention Simms B. T. in Mexico, who in 1946 combined it with rotenone to use it as a spray against arsenic-resistant *Boophilus microplus* strains. He obtained excellent results on young tick forms, but not on undistended females; to destroy these he had to treat the animals with BHC 0.5%.

Nevertheless, in Australia, Hitchcock and Mackerras (1947) obtained an effective control over the same strains using only DDT 1% spray. Blakeslee and Bruce (1948), using a similar method and concentration, obtained the same results, stressing the excellent residual effect and the absence of skin lesions, very common with arsenical compounds.

The latter authors suggested the possibility of controlling the different tick stages off the host, by spraying the fields at a rate of 1.5 kg DDT ha^{-1}, since with this method a 90% control over immature stages can be obtained.

There have been many publications on the use of DDT on various tick species that affect other animals. West and Campbell (1946) described the action of emulsions at 2% for controlling *Rhipicephalus sanguineus* (dog tick), emphasizing the greater susceptibility of the young stages compared with engorged females, which can survive the treatment and lay eggs normally.

Yet Sergent et al. (1945) obtained total control over all those stages using 5% solutions, keeping the dogs free of ticks for 1 week.

DDT Penetration in the Body of Insects

As a contact insecticide, DDT penetrated mainly through the cuticle, particularly in the thinner and more flexible areas.

Tobias et al (1946) point out that, due to its affinity with fats, DDT penetrates easily through the cuticle of insects and even at low concentrations it is sufficient to act by contact. For most insects the LD required for this method of application is even lower than for injections, normally near $10 \ \mathrm{mg \ kg}^{-1}$.

The rapid absorption and subsequent distribution through the hemolymph were studied with external applications of DDT marked with ^{14}C.

Another means of entry is gastrointestinal, where DDT acts as an effective stomach toxic for a large number of insects.

Toxicity for Large Animals

The excellent work done by Negherbon (1959) indicates that, handling the precautions with criterion, toxicity is low compared to that shown by the arsenical compounds. In cattle the LD_{50} is $250 \ \mathrm{mg \ kg}^{-1}$ w/v for acute toxification.

Toxicity does not only depend on the dosage and method of application (oral, respiratory, dermal, placental), but on other factors such as type of excipients, degree of fatness of the animals in oral application.

This substance tends to accumulate in the body fat and in products rich in lipids such as milk and, in the case of birds, in the egg yolk.

The level of DDT in body fat can be high without showing toxic effects. The central nervous system is the principal site of pharmacological activity. The symptoms are: apprehension, excitation, hyperreflexia, ambulatory disturbance, muscle trembling, and finally tonic and clonic convulsions.

Besides lesions in the nervous system, autopsy also shows lesions in the liver; fat hypertrophy and infiltration and, histologically, degeneration of the hepatocite; in the kidney, edema in the cells of the renal tubulae, glomerular congestion, degeneration and necrosis; and focal necrosis in miocardium and voluntary muscles.

Mode of Action of DDT

An increase is observed in the mobility of the ticks affected, followed by contraction of the cutaneous muscles which depress the cuticle in the dorsal and genital grooves.

Later the ticks lose balance, they cannot coordinate their locomotion with or without body movement, and finally their limbs become paralysed.

There are several theories to explain the mechanism of the neurological symptoms shown. Tobias et al. (1946) proved that the effect of DDT on an isolated nerve of an arthropod is to multiply the impulses that pass through the treated section.

Roeder and Weiant (1948) and Welsh and Gordon (1947) maintain that this effect is very similar to that caused by perfusion of fluids with low level of calcium and magnesium ions over an isolated crab axon. The permeability to the calcium ion would be reduced by fixation of DDT to the liquids of the cell membrane.

Symptoms of DDT toxification on cockroaches are similar to those caused by eserine (fisostigmine), a drug whose pharmacological action causes inhibition of the nerve cholinesterase.

Nevertheless, in 1945 Richards and Cutkomp discovered that DDT does not cause cholinesterase inhibition in vitro in an isolated cell of an insect, while Tobías et al. (1946) emphasizes that DDT increases the acetylcholine of the nerve (eserinic effect) not by the enzyme inhibition, but by speeding the liberation of acetylcholine from stored reserves.

Finally Metcalf and Kearns (1945) suggest that the death of insects intoxicated by DDT is caused by exhaustion, as a result of muscular tetany during the final stage of paralysis, and intoxication by the metabolites accumulated in the muscles after the continuous state of contraction.

BHC

Benzene hexachloride was another insecticide compound derived from a benzene ring. From a chemical point of view BHC corresponds to the 1, 2, 3, 4, 5, 6 hexachlorocyclohexane.

Structural Formula:

Chemical Formula: $C_6H_6Cl_6$

Molecular Weight: 290.85

It is synthesized by a simple method based on the chlorination of benzene in the presence of UV light.

The benzene hexachloride thus synthesized consists of a mixture of optical isomers, identified by letters from the Greek alphabet: alpha, beta, gamma, and delta, that have different atomic structures, solubility and melting points. Each isomer is present in proportions of approximately 70, 5, 10 and 15% respectively.

In 1945 Slade and in 1948 Sherman observed that the gamma-isomer of BHC had a considerably greater insecticide effect than the others. For this reason, the toxic capacity of BHC is proportional to its content of the isomer.

This derivative of BHC was named Gamexane by the Imperial Chemical Industry Ltd. At present the purified commercial preparations containing 99% of Gamma BHC are also known by the name of Lindane. As well as an insecticide effect over innumerable species, not mentioned here for being irrelevant to the subject of this book, Gamma BHC also has a definite therapeutic effect on sheep scab, and cattle mange, and against lice on different domestic animals.

After the appearance of arsenical resistance, these synthetic insecticides aroused great interest for the control of different tick species. In 1945 Downing (see Downing et al. 1952) pointed out that Gamma BHC had been successfully introduced into Africa at a minimal concentration of 0.005%, and that the need to control ticks of two and three hosts caused the concentration to be increased to 0.03% with reinforcements of 0.04%.

In 1946 Ault in Argentina studied the effect of this compound for the control of *Boophilus microplus*. He proved that three dips of Gamma BHC at weekly intervals, while harmless to the health of the animals treated, completely cleaned them of ticks.

Formulations

Here we must mention the trials carried out by Whitnall and Bradford (1947), who made several combinations using BHC.

Aqueous Suspensions

The suspensions contained 50% technical hexachlorocyclohexane and were very effective against ticks, causing their gradual death which could take up to 72 days (see Table 5.1).

Wettable Powders

These powders contained 20% technical BHC and needed the addition of water in order to facilitate the mixture in the dipping tank. The normal concentrations should not be below 0.01%, nor over 0.05%.

Table 5.1. Results obtained with aqueous suspensions at different concentrations (Whitnall and Bradford 1947)

Treatment	10 ticks in each treatment		
	Oviposition (%)	Eclosion (%)	Control (%)
Water			
0.001% Gammexane	100	100	0
0.002% Gammexane	30	0	100
0.005% Gammexane	30	0	100
0.01% Gammexane	20	0	100
0.013% Gammexane	10	0	100
0.02% Gammexane	10	0	100
0.025% Gammexane	0	0	100
0.034% Gammexane	10	0	100
0.05% Gammexane	0	0	100
0.067% Gammexane	0	0	100
0.1% Gammexane	0	0	100
0.2% Gammexane	0	0	100
0.25% Gammexane	0	0	100
0.334% Gammexane	0	0	100
0.5% Gammexane	0	0	100
1.0% Gammexane	0	0	100

Table 5.2. Efficacy of gammexane at different concentrations (Whitnall and Bradford 1947)

Treatment	10 ticks in each treatment		
	Oviposition (%)	Eclosion (%)	Control (%)
Caolin	100	100	0
0.01% Gammexane	100	100	0
0.02% Gammexane	100	100	0
0.05% Gammexane	60	66	60
0.1% Gammexane	10	0	100
0.2% Gammexane	0	0	100
0.5% Gammexane	20	0	100
1.0% Gammexane	10	0	100
2.0% Gammexane	0	0	100

By mixing a fine china clay into the active ingredient and varying the concentrations, Whitnall and Bradford obtained the figures shown in Table 5.2. However, these results were not observed in field trials where cattle were treated at weekly intervals, for which reason studies with this formulation were done with less and less frequency.

Norris et al. (1950) carried out trials with Gamma BHC as a dispersable paste with which, on its being mixed into the dip, an 0.025% concentration of gamma-isomer was obtained. It had a very good effect at all tick stages except engorged larvae and engorged nymphs. The protective residual effect was of 8 days, which, as with the ixodicide effect, decreased after the continuous use of the dipping tank.

Gammexane Activity with Other Insecticides

Whitnall and Bradford (1947) carried out comparative studies of the efficacy of Gamma BHC compared with other compounds, stressing its greater efficacy on *Boophilus decoloratus* over arsenic-resistant engorged females. At a dilution of 0.029% a complete control was obtained, while with DDT 1% only a 60% control was obtained (see Table 5.3).

Disadvantages of Gamma BHC

The problems with this product were a consequence of its insolubility in water, due to which it was more easily removable during the passage of the animals, thus causing a loss of biological efficacy. For this reason, some manufacturers suggest making the reinforcement 50% above the initial concentration.

Also, for some BHC formulations, the action in alkaline waters was uncertain, just as in excessively dirty dips, which for this reason should be emptied and cleaned at intervals of not larger than 6 months.

In South Africa, J. Allan (1955) also observed this phenomenon, while in Australia a greater concentration was used in order to avoid it, but without obtaining the expected results.

Several investigators came to the conclusion that the loss of efficacy was due to the action of anaerobic bacteria degrading the active ingredient. A microorganism, called *Calandra canaris*, which was always present in dip washes, was isolated.

Trials were carried out using cultures of this microorganism with dung, urine, scabs and hairs, adding pure BHC isomers. It was proved that the gamma-isomer degraded more rapidly than the rest. Similarly, other experiments were carried out which included a large number of anaerobic microorganisms in peptonated serum, to which the pure gamma-isomer was added. These showed that the bac-

Table 5.3. Comparison of gammexane efficacy with other compounds. (Whitnall and Bradford 1947)

Treatment	50 ticks in water, 10 ticks with each insecticide treatment		
	Oviposition (%)	Eclosion (%)	Control (%)
Water	100	100	0
0.0004% Gammexane	80	100	20
0.0007% Gammexane	80	63	50
0.0015% Gammexane	60	50	70
0.0029% Gammexane	30	0	100
0.0073% Gammexane	0	0	100
0.0145% Gammexane	0	0	100
0.0291% Gammexane	0	0	100
0.0727% Gammexane	0	0	100
0.1455% Gammexane	0	0	100
0.0025% DDT	100	90	10
0.0005% DDT	100	90	10
0.01% DDT	100	100	0
0.02% DDT	100	100	0
0.05% DDT	100	100	0
0.1% DDT	100	90	10
0.2% DDT	100	90	10
0.4% DDT	70	86	40
1.0% DDT	60	66	60
0.16% AS_2O_3	100	60	40
0.32% AS_2O_3	100	60	40
0.64% AS_2O_3	60	0	100
1.28% AS_2O_3	40	0	100

terial development was accompanied by a substantial reduction of BHC. As regards the hydrogen produced by bacteria, the phenomenon was chemically interpreted as acting in dechlorinating the compound.

Penetration of BHC

Armstrong et al. (1951) studied the penetration of BHC isomers through the insect cuticle using microcolorimetric techniques, exposing the wheat worm on filter paper impregnated with the isomers.

It was established that the quantities of each isomer picked up by the insects were determined by two fractions: an external one, which could be removed by washing with methanol; and an internal one, recovered after decomposition of the insect with nitric acid.

The results showed that the gamma-isomer penetrated through the layers of insect tegument faster than the other isomers. The author explained this phenomenon by the molecular configuration with respect to the cuticle structure, that played an important part in the passage.

However, these investigations did not coincide with the hypothesis of Dresden and Krijgsman (1948), who were of the opinion that the toxicity of the gamma-isomer was not a consequence of the rapid penetration, but that it was due to that particular molecular configuration allowing the attack by inknown physiological processes. They proved the theory by infecting *Periplaneta americana* cockroaches with the four isomers: alpha, beta, gamma and delta, and showed that the gamma-isomer was the most toxic.

Symptoms of Intoxication in Insects

Boophilus microplus. Ault (1948) points out that already at dilutions of 1:2500, ticks suffer severe contractions that prevent locomotion. Whitnall and Bradford (1947), confirmed these symptoms, saying that after 12 h of treatment the ticks appear flattened dorsoventrally, with the anterior and posterior ends turned inwards. Ticks do not die immediately, but gradually, starting 9 h after dipping, sometimes showing slight movements of their limbs up to 60 days later, and finally dying.

Pasquier (1947) describes similar findings in locusts (*Squistocerca* spp. intoxicated by Gamma BHC, where the symptoms consisted of changes in the abdomen, hyperexcitability to stimulations, attempts at flying, incoordination and paralysis.

Mechanism of Action

The mechanism of action of organochloride compounds is not fully explained. Slade (1945) suggests that BHC acts by blocking a vital function of the ticks. This would be due to Lindane having a structure similar to that of inositol, which it would inhibit, by not allowing it to fulfill its specific function, essential for the ixodide, thus substituting it and causing a blockade in the metabolism.

This affirmation is shared by several investigators, and is based on experiments developed by Kirkwood and Phillips (1946) with the yeast denominated *Saccharomyces cerevisiae*, which requires inositol for its development. They proved that Lindane inhibits the development of the above-mentioned strain, while the other isomers are relatively inactive. The effect of Lindane can be annulled by adding mesoinositol.

Other authors attribute the toxicity of chlorinated insecticides to the freeing of hydrochloric acid which, according to Pacheco Torres (1968), would be the main characteristic of DDT, BHC, chlordane and toxaphene.

An important point to emphasize is that the main action is done by contact and, secondly, by ingestion through the mouth parts of the insects. This statement is based on studies carried out by Thorburn (1968) who, occluding the mouth area of ticks with high-concentration Vaseline, compared the time they took to die with that taken by ticks that could ingest freely. He observed that there was almost no difference.

Another point worth noting is that Lindane acts on insects mainly as a neurotoxin on the nervous ganglion, unlike DDT, which acts on the axon. For this reason there can never be cross-resistance between Lindane and DDT.

Residual Effect

The duration of the residual effect is in direct relation to the concentration, oscillating around 12 days.

As Purchase (1955) points out, despite the excellent effect of the active ingredient in arsenic-resistant strains, after a short period of approximately 18 months, according to Wilson (1978) the ticks also started to show tolerance to BHC in South Africa and Australia, where the fight had started so successfully.

Yet as Newton (1967) points out, their use for the control of ixodides continued until 1962 in formulations with mixtures of DDT, due to the restrictions placed on organochloride compounds.

From then on, Gamma BHC was only used for the control of other types of acarus, due to the enormous residues left in the animal tissue, as is shown in studies on the metabolism of this active ingredient, carried out by Clark et al. (1956).

Toxaphene

Chemical Name: Octachlorcamphene

Structural Formula:

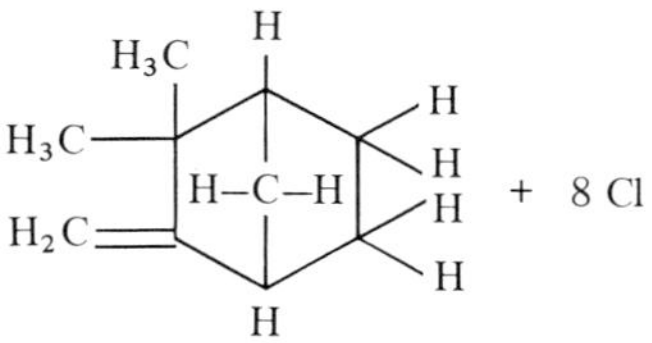

$$C_{10}H_{16} + 8\,Cl$$

Chemical Formula: $C_{10}H_{16}Cl_8$

This active ingredient was widely used in South Africa (Purchase 1955) for the control of various tick species such as *Amblyomma, Hyalomma* and *Rhipicephalus* on various hosts. It showed a greater ixodicide effect than the other organochlorides.

Toxaphene is insoluble in water, but highly soluble in organic solvents like xylol, and in miscible oils, with which it forms very stable emulsions.

The most efficient concentrations are those between 0.3 and 0.7%.

Owing to its resinous nature, this substance adheres very well to the skin and hairs of the animals treated, showing identical properties when used by immersion or aspersion, as well as having a prolonged residual effect.

This effect is independent of the volume used. Drummond et al. 1966 showed that in cattle infested with *Amblyomma americanum* and *Dermacentor albipictus*, in which for one lot 3.7 l per animal were used and for the other 1.5 l; the same level of reinfestation was observed after 3 weeks of treatment. Due to the important background of efficacy on other species, its use in Argentina started in 1948. An excellent description of its action against *Boophilus microplus* is found in a publication by Ault (1950). He carried out comparative studies at 0.47, 0.59 and 0.7% in only one plunge dip, pointing out the prolonged residual effect of 12 days and the action over the different stages of the life cycle which we transcribe below:

Engorged females: rapid dropping off which is completed in 48 h, annulling spawning completely.

Semi-engorged females; undistended females and males: 100% lethal.

Engorged nymphs: the least affected, even though only 38% completed their life cycle.

Engorged larvae: irregular effect, although the nymphs emerging from these larvae later died as a result of the residual effect.

Wood (1967) points out the simplicity of the method of analyis to establish the concentration of active ingredient of the dipwash samples and the low cost of toxaphene, as well as other features that facilitate its use. He also points out that no other compound, except chlorfenvinfos, is comparable to toxaphene in tick control on more than one host. Notwithstanding the evident excellent action over BHC-resistant strains in East Africa, when it was used repeatedly at short intervals (Barnett 1961), there were toxaphene-tolerating *Boophilus* in most places where there was resistance to organochloride compounds.

Chlordane

Chemical Name: Octachlorodihydrocyclopentadiene, octachloro, or 1:2:3:4:5:6:7:10:10-octachloro-4:7:8:9-tetrahydro-4:7-endomethylene indane.

Structural Formula:

$$\text{Cl, Cl, Cl, Cl (on CCl}_2\text{ ring); H, H, Cl, H, Cl, H}_2\text{ substituents}$$

Chemical Formula: $C_{10}H_6Cl_8$

With chemical properties similar to those of toxaphene, chlordane showed a great penetration capacity which proved very effective against insects with a very sclerotyzed cuticle, as is the case with locusts. It was not used very extensively, and its action proved to be similar to that of toxaphene.

Experiments were developed in Australia with an emulsion of chlordane 0.5%, which offered a 99% control over *Boophilus* and had a protective residual effect of 7 days. At 0.25% it offered an 84% control and protected against reinfestation during 6 days.

In America, Brundrett et al. (1955) tried an emulsion at 0.5% against *A. Americanum*. The ticks dropped off within 48 h and there was protection up to 5 days post-treatment.

Dieldrin

Chemical Name: 1,2,3,4,10,10-Hexachloro-6,7-epoxy-1,4,4a,5,6,7,8,8a-octa-hydro-1,4-endo,exo-5,8-dimethanonaphthalene.

Structural Formula:

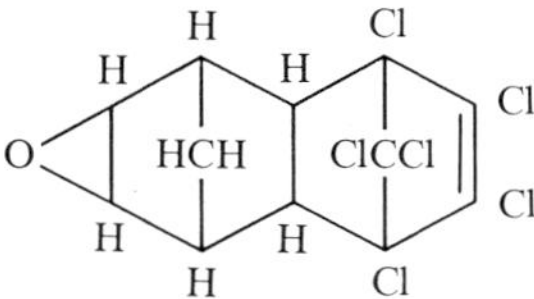

Chemical Formula: $C_{12}H_8Cl_6O$

Like other chlorinated hydrocarbons, dieldrin is insoluble in water and soluble in organic solvents, although to a lesser degree than aldrin, with a residual effect similar to DDT. Its toxicity is also smaller than that of other chlorinated insecticides. It has a systemic activity when applied subcutaneously at dosis of $25\ mg\ kg^{-1}$ w/v, and lasts 43 days, not allowing larvae infestation (Roulston 1956).

The first trials on cattle infested with *Boophilus microplus* were carried out by Legg (1953). He sprayed with a concentration of 0.1%. The trials showed the excellent activity of this active ingredient during the first 24 h, over all the young stages. At half this concentration there are also references on its efficacy over most of the stages a few hours after application.

Barnett (1961) points out that at 0.05% spraying at regular intervals, it provides results comparable with toxaphene and chlordane. Nevertheless, its use never became widespread, for its appearance coincided with the loss of efficacy of BHC, and due to the rapid development of resistance of the treated strains.

In this aspect, work done by Stone and Meyers (1957) deserves special consideration; they recorded the high level of resistance offered by a *Boophilus microplus* strain when treated by aspersion with 0.05% emulsions.

Aldrin

Chemical Name: (1,2,3,4,10,10a-hexachlor-1,4,4a,5,8,8a-hexahydro-1,4-endo, exo-5,8-dimethanonaphthalene)

Structural Formula:

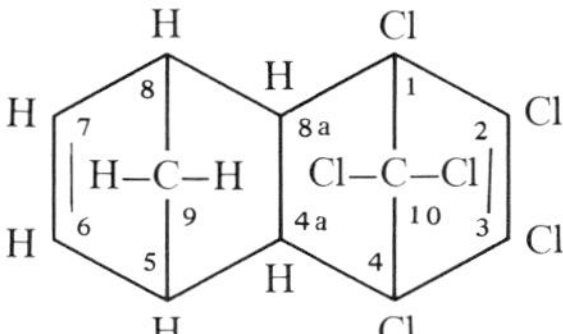

Chemical Formula: $C_{12}H_8Cl_6$

It is a white crystalline solid, less volatile than lindane, with a residual effect lying between that of chlordane and of DDT, of approximately 5 days. Like all the rest of the group, aldrin is insoluble in water, soluble in organic solvents and very stable to alkali.

Its properties are similar to those of dieldrin and its toxicity in insects equivalent to that of lindane.

Roulston and Wilson (1965) carried out comparative trials by aspersion at concentrations of 0.05 and 0.01% in cattle, obtaining more conclusive results with 0.01%.

Roulston (1956) independently studied the systemic action of aldrin, applying dosis of 25 mg kg^{-1} w/v subcutaneously. He obtained a marked effect on cattle heavily infested with *Boophilus microplus.* He also observed a strong residual effect over larvae reinfestations up to 80 days after application.

Regarding its behaviour over BHC-resistant strains, like dieldrin, aldrin shows no effect, as with toxaphene and chlordane (Norris 1956).

Organophosphorous Compounds

In order to avoid on the one hand residues in the tissues of the animals treated, and on the other the problems related to the resistance to chlorides by ticks, the search for new active ingredients is continuous. The result of this search was the appearance of organophosphorous compounds, which at the time were considered the most effective ixodicides.

All the components of this group are organic preparations derived from phosphoric acid, for which reason they have been given the common name of organophosphorous compounds, being also characterized by a similar mode of action (Barnett 1961).

Of their qualities, we can mention the fact that they do not leave residues in animals for slaughter, an enormous advantage over the group of ixodicides that

preceded them. Clarborn (1956) sprayed a group of cattle with malathion 0.5%
16 times at weekly intervals. A week after treatment he did not find any traces in
the fat of the animals treated.

These results were later demonstrated by March (1956), who also carried out
trials in cattle with malathion, marking it chemically with phosphorous 32 and
finding minimal residues in the meat, around 0.05 and 0.15 ppm.

In milk from cattle treated by aspersion twice in 1 week with doses of 0.3%
and 0.1%, only 0.08 to 0.36 ppm were discovered 5 h after the last application,
and after 24 h, some traces which disappeared completely 3 days later.

Physiopathology

The toxic action of this group of compounds is based on the inhibiting effect of
acetylcholinesterase enzyme resulting in an increase of acetylcholine, which on
rising to critical levels causes the appearance of the symptoms and a posteriori a
lethal effect.

As will be recalled, in mammals the axon ends in a neural plaque situated in
the muscle fibre or in the organ called neuromuscular, which is activated by
acetylcholine.

The sinapsis of pre- and post-ganglionic fibres of the parasympathetic
system, as also the preganglionic of the sympathetic system, is measured by
acetylcholine, for which reason they are named cholinergics, while the post-ganglionic fibres of the sympathetic use adrenalin and noradrenalin and are called
adrenergics (O'Brien 1960 (see Fig. 5.8).

The transmission of the impulse is produced when the acetylcholine is
liberated from the vesicles that contain it (Guyton 1963), originating the stimulus
in the corresponding efferent organ.

If the acetylcholine persisted, it would continue stimulating indefinitely. In
practice this does not occur because it is immediately destroyed by the acetylcholinesterase, present in large quantities. If this did not act, then the acetylcholine could accumulate, causing first an excess of stimulation and finally the
blocking of the cholinergic system.

As a result of the cholinesterase inhibition, there would be an interference at
the point of neuromuscular joint, increasing the contraction of voluntary
muscles up to paralysis.

Function of Acetylcholine

Acetylcholine has two fundamental functions:
1. Muscarinic or pure parasympathetic mimetic with action on glands, and innervated smooth muscles by parasympathetic postganglionic fibres.

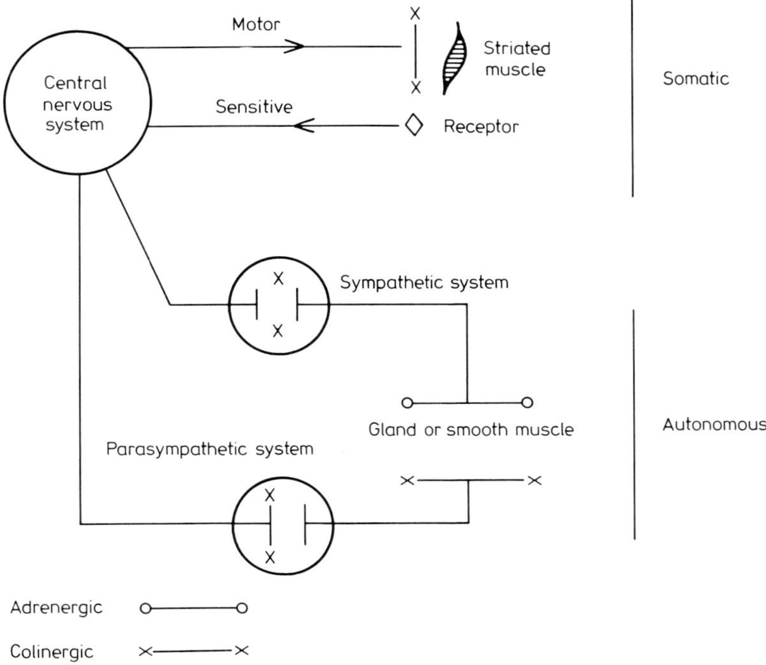

Fig. 5.8. Major subdivisions of the nervous system

2. First nicotinic of stimulation and then paralysis over ganglions of the autonomic nervous system, both sympathetic and parasympathetic, of the voluntary muscles.

Generally these effects are antagonic, the first ones predominating.

Acetylcholine is found in the nervous system, in the glands, in the muscles, and incidentally in the erythrocytes. It originates from the combination of acetic acid and choline amino acid catalyzed by the cholinesterase enzyme. Figures 5.9 and 5.10 will help explain this.

Acetylcholine Breakdown by Cholinesterase

As said above, under normal conditions acetylcholine is hydrolysed immediately it is produced at the level of nervous synapses by an enzyme, acetylcholinesterase, which acts upon the acetylcholine at a low concentration and upon certain synthetic esters of choline which have the nervous sinapse as fundamental localization.

There is another unspecific enzyme called pseudocholinesterase that has the property of acting upon actylcholine at a high concentration, hydrolysing almost

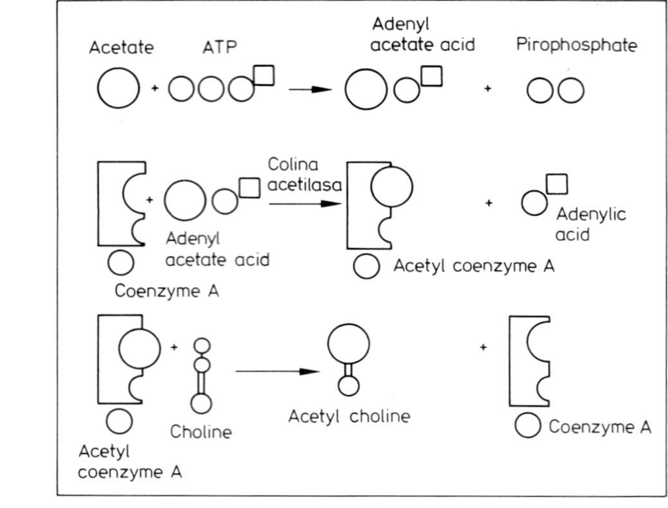

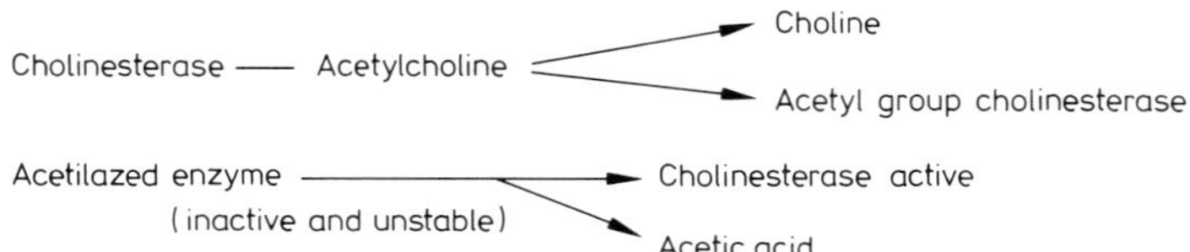

Fig. 5.9. Synthesis of acetylcholine

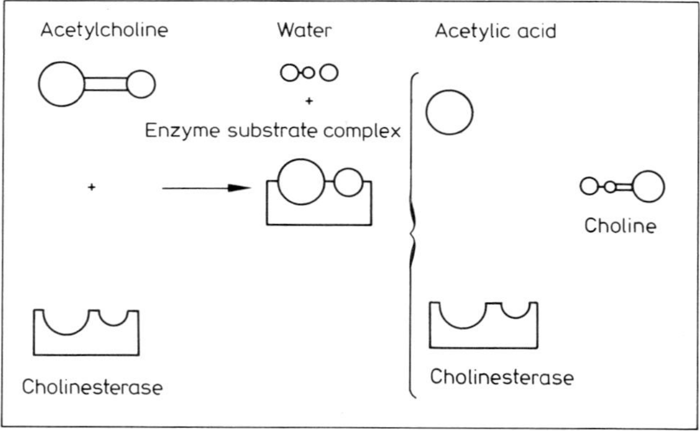

Fig. 5.10. Breakdown

all the esters of the choline. It is found in blood plasma and in different tissues, including CNS.

Under physiological conditions, the cholinesterase acts upon the acetylcholine, liberating the choline amino acid and combining it with acetyl group, thus constituting an acetylated enzyme in the form of an inactive, very unstable compound, which causes it to break down rapidly into acetic acid and active cholinesterase, ready to act once again.

Cholinesterase Active Centres and Breakdown of Acetylcholine

The enzyme may have two active centres, one anionic nucleus and another estearic cationic (see Fig. 5.11). The former acts by attracting the positive charges of the quaternary nitrogen atoms of the acetylcholine, thus fixing and orientating the substrate and so the stearic nucleus, which is the really active one, carries out its hydrolitic action.

The action of organophosphorous compounds is probably due to the phosphorylation of the stearic group cholinesterase, forming a combination which, unlike the acetylated enzyme, is very stable. Thus the enzyme activity lasts a long time, allowing the level of acetylcholine to rise, and determine the appearance of toxic symptoms.

Thus we notice that the organophosphorous compounds produce their toxic effect, for they act like anticholinesterase, adhering to the enzyme irreversible (see Fig. 5.12).

In ticks the nervous transmission would be regulated by an analogous system of chemical mediators, and consequently the toxic effect of the organophosphorous compounds would be similar.

This is how Roulston et al. (1965) saw it when studying the metabolism of coumaphos in tick larvae, and Stone (1968) in the nerve tissues in adult stages.

In other experiments, Roulston et al. (1968a) correlated the signs of intoxication in ticks with the degree of cholinesterase inhibition.

The resistance of some strains is also related to the degree of inhibiting action of the acetylcholinesterase enzyme (Wharton 1967) (see corresponding chapter).

On the basis of these theories, from the beginning the presence of acetylcholine was presumed, as well as the enzyme that intervenes in its synthesis and

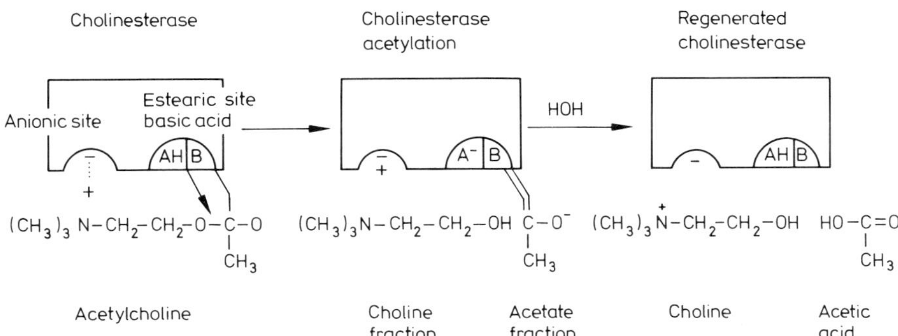

Fig. 5.11. Active centres of cholinesterase and degradation of the acetylcholine

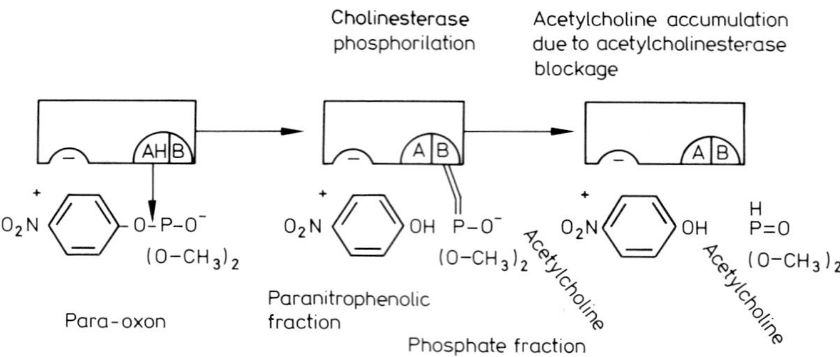

Fig. 5.12. Blockage of the cholinesterase by an organophosphorate

the choline acetyltransferase which were experimentally confirmed by Smallman and Schunter (1972). Using organic solvents, these authors were able to obtain extracts of a biological substance which, when subjected to electrophoresis, migrated exactly like acetylcholine. Subsequently, they were able to confirm the identity of this biologically active substance by characterizing it with chromatography and radioactive [14]C. On the other hand, using these extracts on abdominal rectum muscle tissues of frogs and of the fundic region of rat stomach, they were able to abolish the reaction to stimulus after having incubated the above-mentioned substance with the acetylcholinesterase enzyme.

In another part of this experiment they used brain extracts of several hundred adult ticks, identifying by electrophoresis the biologically active substances such as acetylcholine, present in the proportions of $47\,\mathrm{g}^{-1}$; a considerably larger amount than that contained in larvae, which was $12.6\,\mathrm{g}^{-1}$.

According to their results, these experiments serve to confirm the presence of the cholinergic system and reply to the hypothesis of the lethal action of organophosphorous compounds in *Boophilus microplus*.

Entrance into the Body of the Arthropod

As efficient contact insecticides the organophosphorous compounds penetrate rapidly through the cuticle.

Indeed, for some of them, like parathion, it is equally toxic to apply them by contact or by injection, according to studies carried out on *Periplaneta americana*.

Nevertheless, differences were detected in the speed of action to obtain the knockdown of this species, depending on the proximity of the central nervous system to the point of contact.

Another factor that can vary the speed of entrance or passage of the active ingredient through the cuticle is the type of solvent used in the formulation.

Distribution

With studies carried out with marked radioactive compounds, it was possible to determine that the distribution of toxic compounds in the insects is done via the circulatory system (Metcalf 1955).

Symptoms of Intoxication by Poisoning in Ticks

These are manifested by hyperexcitability and hyperactivity developed almost immediately after the injection, followed by muscle incoordination, tonic clonic convulsions and death.

Organophosphorous Toxicity in Mammals

Even before the appearance of the clinical symptoms of intoxication, it is feasible to find a decrease of cholinesterase in the blood. This enzyme can be investigated by the simple method described by Jolly and Ratcliffe (1958), who also suggest the periodic use of this test in areas where organophosphorous dips are used regularly.

The test is based on the titration of the acetic acid produced by the interaction of the acetylcholine and the cholinesterase from a blood sample. The titration is done with the aqueous solution of bromothymol as indicator.

Symptoms of Intoxication in Mammals

The symptoms are caused by the cholinesterase inhibition in different parts of the body, including post ganglionic, cholinergic nerves (muscarinic effect), preganglionic and somatic motor nerves (nicotinic effect) and of the central nervous system.

The muscarinic effects are: anorexia, abdominal pains, salivation, profuse perspiration, miosis; in more severe cases: diarrhea, tenesmus, involuntary defecation and urination, excessive bronchial excretion, respiratory difficulties, lung edema with cyanosis, pronounced oliguresis, coma and death.

Necropsia Findings

The *postmortem* lesions are common to other intoxications and are not pathognomonic (Radeleff 1958).

Intoxication Treatment

In cattle it consists basically in administrating atropine sulphate in doses of 0.5 mg kg^{-1}.

It is advisable to apply part of the dose intravenously. Woodward (1957) proposes the use of 2 pam (piridinealdoxinemethiodide) together with the atropine.

Commercial Formulations

Commercial Name: Nexagan

Laboratory: Shell

Chemical Name: 4-Bromo-2,5-dichlorophenyl diethyl phosphorothioate

Structural Formula:

$$
\begin{array}{c}
C_2H_5O \\
| \\
C_2H_5O-\overset{\displaystyle}{\underset{\underset{S}{\|}}{P}}-O-
\end{array}
\text{—phenyl ring with Cl (top), Br (right), Cl (bottom)}
$$

Chemical Formula: $C_{10}H_{12}O_3Cl_2BrSP$

Solubility: Relatively insoluble in water

Formulation: 40% Concentrate emulsifiable with 5% chlorfenvinphos

Recommended Dosage: Dipwash: 0.12%; replenishment and reinforcement: 0.15%

Toxicity: Acute oral LD_{50} (rats): 55 mg kg^{-1}
Acute dermic LD_{50} (rabbits): 1366 mg kg^{-1}
LD_{50} in cattle: no data available.

The acaricide properties of this compound were first established in South Africa (Fiedler and Vuuren 1966) with trials in vitro carried out with engored females of *Boophilus microplus*, at a concentration of 200 ppm. Nexagan also proved to be effective in the fight against different tick species of more than one host when used at 0.05% active ingredient or at 0.03% when applied with toxaphene 0.25% (Drummond et al. 1968).

The compound also has an excellent effect on *Hipoderma spp.* (Drummond 1966). Trials carried out by CSIRO (1967) and Roulston and Wharton (1967)

show the excellent results of bromophos ethyl in strains resistant to organophosphorous compounds such as Biarra, Brisbane and Ridgelands.

Although they later detected a few factors of resistance to this compound, Shaw et al. (1968) observed a good level of activity against the same strain.

Owing to the versatility of the formulations, Nexagan was also evaluated as a pour-on, as indicated in work done by Loomis et al. (1972), who obtained an efficacy over *Boophilus microplus* varying between 60% and 90%.

In Argentina Grillo Torrado et al. (1971) worked in vitro and in vivo, and pointed out the high oviposition inhibition of Bromophos ethylic 40% mixed with chlorfenvinphos 5%. Mc Cullough and Nolan (1974) emphasized that the said mixture showed the same efficacy as ethion over *Boophilus microplus*.

Carbonphenothion

Commercial Name: Garration

Laboratory: Staufer

Chemical Name: S-(4-Chlorophenylthiomethyl) diethyl phosphorotiolothionate

Structural Formula:

$$C_2H_5O-\underset{\underset{S}{\overset{\|}{}}}{\overset{\overset{C_2H_5O}{\overset{|}{}}}{P}}-S-CH_2-S-C_6H_4-Cl$$

Chemical Formula: $C_{11}H_{16}O_2PS_3Cl$

Solubility: Very insoluble in water

Excretion: No data available

Formulation: 20% w/v, Emulsifiable concentrate

Recommended Dosage: 0.02% w/v

Toxicity: Acute oral LD_{50} (rats): 32 mg kg^{-1}
Acute dermic LD_{50} (rabbits): 127 mg kg^{-1}
LD_{50} in cattle: no data available.

When this product was used, its balanced formulation proved very stable in dips under field conditions and was highly emulsifiable. However, as Boero (1969) pointed out, its ixodicide activity was slower than that of ethion.

Chlorfenvinphos

This is a non-systemic insecticide and acaricide. It was widely used with bromophos ethyl as a component of Nexagan.

Commercial Name: Supona

Laboratory: Shell

Chemical Name: 2 Chloro-1-(2,4-dichlorophenyl)vinyl diethyl phosphate

Structural Formula:

$$C_2H_5O-\underset{\underset{O}{\|}}{\overset{\overset{C_2H_5O}{|}}{P}}-O-\underset{\underset{CHCl}{\|}}{C}-C_6H_3Cl$$

Chemical Formula: $C_{12}H_{15}O_4PCl_2$

Solubility: Good in organic solvents and scarce in water

Formulation: Mixed (5%) with bromophos ethyl (40%) emulsifiable concentrate

Recommended Dosage: Dipwash: 0.04%; replenishment and reinforcement: 0.15%

Toxicity: Acute oral LD_{50} in rats: $10-39$ mg kg^{-1}

Acute dermic LD_{50} in rats: $31-108$ mg kg^{-1}

Acute oral LD_{50} in cattle: 20 mg kg^{-1}.

This organophosphorous compound, which had a very good background as an insecticide for lice and flies of different species, began being developed after work carried out by Graham and Drummond (1964), who confirmed its effect against ticks of the *Boophilus* species.

It was later studied extensively on different species, with acceptable results. Shaw and Baker (1966) found that chlorfenvinphos had a better effect on adult stages than on larvae, both in *B. microplus* and in *B. decoloratus.*

J. A. F. Baker and Thompson (1966) worked with concentrations of 0.01, 0.02 and 0.05% in field trials carried out in South Africa. He obtained similar results in both mature and young stages in tick species of various hosts. Similarly, at 0.05% concentration, Thorburn (1968) suggested the use of this active ingredient for the complex control of the same species.

Chlorfenvinphos was used actively in Mexico at a 20% concentration (Jara de la 1971), while in Argentina it was formulated at 40%.

Chlorpyriphos

Commercial Name: Dursban

Laboratory: Dow

Chemical Name: 0,0-Diethyl-0-(3,5,6-trichloro-2-pyridyl) phosphorothioate

Structural Formula:

$$C_2H_5O-\underset{\underset{S}{\|}}{\overset{\overset{C_2H_5O}{|}}{P}}-O-\text{(3,5,6-trichloropyridyl)}$$

Chemical Formula: $C_9H_{11}O_3NSCl_3P$

Solubility: Scarce in water

Formulation: Approximately 25% w/v

Recommended Dosage: Dipwash: 0.028%; replenishment and reinforcement: 0.049%

Toxicity: Acute oral LD_{50} in rats: $97 - 276$ mg kg^{-1}
Acute dermic LD_{50} in rabbits: 2000 mg kg^{-1}
LD_{50} in cattle: unknown.

Kenaga et al. (1965) describe for the first time the excellent insecticide effect of the compound; Whitney et al. (1967) found the same efficacy on cockroaches, and Ludwig and Mc Neill (1966) used it with success in the control of mosquitoes.

Coollier and Dieter (1965) confirmed these results for the control of other insects, while in *Boophilus microplus* Wade (1968) carried out large-scale field trials, emphasizing the efficacy and stability of the active ingredient when used at an initial concentration of 0.05%, without detecting signs of toxicity to the animals.

Danilo Parra (1969) points out that even at concentrations of 0.0125%, chlorpyriphos is efficient over all the sensitive stages of *Boophilus microplus*, also observing that it had a residual effect of 2 weeks.

However, in Argentina, even using it at double the concentration, it was not possible to obtain these results, particularly in semi-engorged females (Grillo Torrado and Gutierrez 1977).

Coumaphos

Commercial Name: Asuntol

Laboratory: Bayer

Chemical Name: 0,0-Diethyl-0-(3-chloro-4-methyl-7-coumarinyl) phosphorothioate

Structural Formula:

C_2H_5O and C_2H_5O attached to P, with P–O bond to ring; P=S; ring system with O, =O, Cl, and CH_3 substituents.

Chemical Formula: $C_{15}H_{16}O_5ClPS$

Solubility: Practically insoluble

Formulation: 20% w/v Emulsifiable liquid; 50% w/w wettable powder. This product was used in two different formulations

Recommended Dosage: Emulsifiable liquid: dipwash = 0.03%; replenishment and reinforcement = 0.06%
Wettable powder: Dipwash = 0.08%, replenishment and reinforcement = 0.10%

Toxicity: Acute oral LD_{50} in rats: 61.5 mg kg^{-1}
Acute dermic LD_{50} in rabbits: low
LD_{50} in cattle: no data available.

Laboratory and field studies with coumaphos carried out by Fiedler and Veldman (1957) showed its biological efficacy on different tick species such as *Amblyomma hebraeum, Boophilus microplus, Hyalomma* spp. and *Rhipicephalus appendiculatus.*

In 1959 and 1960, Drummond et al. confirmed these at different concentrations. Drummond rejected the systemic action (1958), as he obtained no effect when it was applied subcutaneously and orally to guinea pigs infested with *Amblyomma americanum.* Neither did Vickery and Arthur (1960) obtain results in rabbits infested with the same species.

Stone and Webber (1960) observed a positive ixodicide effect of coumaphos in *Boophilus microplus* resistant to DDT. Regarding the reproduction of this species, Graham and Drummond (1964) treated engorged females with concentrations of 0.01 and 0.1%, and succeeded in annulling oviposition completely. In other trials, taking the same stages, but with *B. annulatus*, in cattle treated with coumaphos 0.25%, although without annulling the reproduction, Graham et al. (1964) observed the total inhibition of oviposition.

Also Drummond et al. (1964), treating cattle infested with *Boophilus microplus* and *Boophilus annulatus*, observed between 92% and 100% lethality in engorged females and 67 – 100% inhibition of shedding engorged nymphs, as well as complete inhibition of oviposition.

Regarding the protective residual effect, work done by Jung (1959) is interesting. He establishes almost the same length of time as for organochloride compounds, that is 9 – 10 days.

Roulston et al. (1965) studied the metabolism of coumaphos in *Boophilus microplus* larvae, using chromatography and spectrography. In these studies it was observed that during metabolism an analogous oxygenated compound was formed, which in vitro had a greater acetylcholinesterase inhibition than its predecessor, as well as other metabolites soluble in water with less lethal effect.

The most commonly used formulations were the wettable powders of 30 to 50% and the emulsifiable concentrate of 16 to 20% active ingredient. Said et al. (1971) obtained with both these formulations the same results on mortality of engorged females, oviposition and viability of *Boophilus microplus* eggs.

Diazinon

Commercial Name: Neocidol

Laboratory: Ciba-Geigy

Chemical Name: 0,0-Diethyl-0-(2-isopropyl-4-methyl-6-pyrimidil) phosphothioate

Structural Formula:

$$C_2H_5O-\overset{\displaystyle C_2H_5O}{\underset{\displaystyle S}{\overset{|}{\underset{\|}{P}}}}-O-\text{(pyrimidyl ring)}$$

Chemical Formula: $C_{12}H_{12}O_3N_2SP$

Solubility: Normal in organic solvents

Formulation: 60%

Recommended Dosage: Dipwash = 0.03%; Replenishment and reinforcement = 0.045%

Toxicity: Oral LD_{50} in rats: 850 mg kg^{-1}
Acute dermic LD_{50} in rats: 2150 mg kg^{-1}
LD_{50} in cattle: no data available.

This compound was discovered in 1951 by Dr. Gysin in Ciba-Geigy laboratory in Basel (Switzerland); this firm was in charge of its subsequent development. Diazinon also showed an effect over other ectoparasites (Grannet and Shea 1955) and insects (Kirshnan et al. 1957), and proved harmless to animals (Legg 1956) and the operator. Unlike the chlorides, it has the advantage of not leaving unacceptable residues in the animal tissues, that could interfere with the normal marketing of meat.

Blin and Gunther (1955) carried out different studies on residues in milk from cows kept in stables treated with the compound, finding only traces of it.

Regarding its action against *Boophilus microplus* Legg (1956) recommended the use of concentrations not below 0.05% active ingredient. Although it was very efficient over undistended females, semi-engorged nymphs and larvae, over engorged nymphs and engorged females the effect decreased ostensibly. Roulston and Wharton (1967) points out that the compound proved inefficient against strains resistant to certain organophosphorous compounds in Queensland.

After carrying out several plunge dip trials, dipping 6000 head of cattle in Argentina, Pizarro et al. (1963) observed a fault in the control of *Boophilus microplus* when they detected live engorged nymphs, after two dips, of undistended females, semi-engorged females and some engorged females.

The author also pointed out that the compound did not affect the fertility of eggs, estimating an 85% hatching, and neither did it reduce or damage the vitality of the larvae hatched, establishing a residual effect of only 1 day.

Ethion

Commercial Name: Coopertox E8

Laboratory: FMC − Cooper Argentina

Chemical Name: 0,0,0′,0-Tetraethyl-S,S-methylene-biphosphorodithioate

Structural Formula:

$$\begin{array}{ccc}
OC_2H_5 & & OC_2H_5 \\
| & & | \\
C_2H_5O-P-S-CH_2-S-P-OC_2H_5 \\
\| & & \| \\
S & & S
\end{array}$$

Chemical Formula: $C_9H_{22}O_4P_2S_4$

Solubility: Slightly soluble in water and soluble in most organic solvents

Formulation: Concentration of active drug = 96% w/v

Recommended Dosage: Dipwash = 0.120%; replenishment and reinforcement = 0.192%

Toxicity: Acute oral LD_{50} in rats: 24.4 mg kg^{-1}
Acute dermic LD_{50} in rabbits: 915 mg kg^{-1}
LD_{50} in cattle: no data available.

As seen in trials carried out by Roulston and Wilson (1965), Drummond et al. (1967), and Palmer (1976a), this compound proved to be active against *Boophilus microplus* at 0.075%. Also, it has a high percentage of efficacy against stages of the life cycle named "resistant", such as engorged nymph. These stages, being protected by the cuticle of the preceding stage, offer a greater resistance to the penetration of the active ingredient (Grillo Torrado and Gutierrez 1977). It proved to be a solution for the control of strains resistant to other organophosphorous compounds, as was described in susceptibility tests with 20 strains from the campaign area developed by Nuñez et al. (1972) in Argentina. These showed only slight signs of tolerance, thus providing the animal health authorities with an efficient weapon for use in their sanitary campaign.

Work done by Shaw (1965) also points out the fact that within the organophosphorous compounds ethion is very effective against resistant strains.

According to Roe (1969), due to these attributes, prior to 1972, 1500 dipping tanks in the quarantine area in New South Wales were filled with ethion. It is mentioned that it also caused a correct inhibition over oviposition (Grillo Torrado and Perez Arrieta 1973). Its formulation is very stable in immersion dips submitted to the passage of large numbers of cattle. Regarding this aspect we can mention work carried out by Nuñez et al. (1975) who treated 50,324 animals over 675 days, maintaining the concentration of active ingredient between 0.09 and 0.13%, with an efficient therapeutic level that cleaned many herds destined for areas free of the parasite.

Without doubt the plunge dip proves more effective than the spray: Wharton et al. (1970) obtained almost 9% more efficacy.

In conclusion we can say that in the more tolerant strains Palmer (1976b) was able to increase the efficacy by mixing the compound with chlordimeform.

Phosmet

Commercial Name: Imidan

Laboratory: Stauffer — Ciba-Geigy

Chemical Name: 0,0-Dimethyl (S-phtalimidomethyl) phosphorodithioate

Structural Formula:

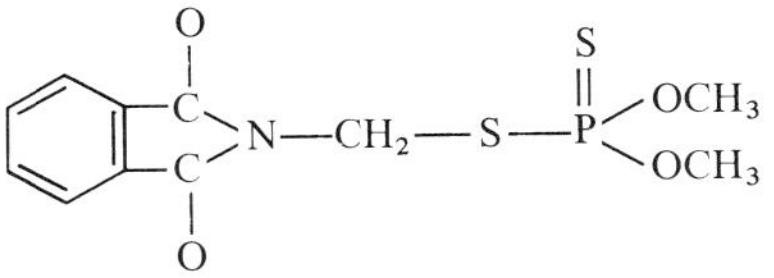

Chemical Formula: $C_{11}H_{12}O_4NS_2P$

Solubility: Fair in organic solvents

Formulation: 15% Emulsifiable concentrate

Recommended Dosage: 0.075% w/v

Toxicity: Acute oral LD_{50} in rats: $230 - 299$ mg kg^{-1}
Acute dermic LD_{50} in rabbits: 3160 mg kg^{-1}
LD_{50} in cattle: no data available.

This compound showed a relative efficacy against the tick strain named Biarra (Roulston and Wharton 1968). It was not marketed until 1970 due to the residues detected in meat.

Dioxathion

Commercial Name: Delnav

Laboratory: Hercules

Chemical Name: S,S-1,4-Dioxane-0,2,2-diyl-bis-(0,0-diethyl-phosphorodithioate)

Structural Formula:

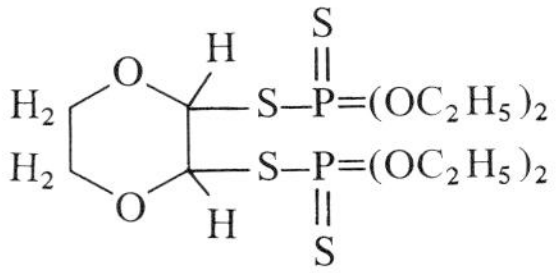

Chemical Formula: $C_{12}H_{26}O_6S_4P_2$

Solubility: Insoluble in water, but soluble in organic solvents

Formulation: 60%

Recommended Dosage: Dipwash = 0.15%; replenishment and reinforcement = 0.20%

Toxicity: Acute oral LD_{50} in rats: 43 mg kg^{-1}
Acute dermic LD_{50} in rats: 235 mg kg^{-1}
LD_{50} in cattle: no data available.

This compound was introduced as from 1954 as an acaricide, and was widely used in agriculture.

The preliminary trials on ticks were carried out in Brazil in 1955, with herds infested with *Boophilus microplus* strains resistant to toxaphene (Cooper 1962).

During 1957 it was successfully used in South Africa at different concentrations and on different species such as *Rhipicephalus appendiculatus, Boophilus decoloratus* and *Rhipicephalus eversti.*

In order to improve its efficacy against these species, J. A. F. Baker et al. (1962) carried out trials mixing dioxathion with 0.025% chlorfenvinphos and obtained better results than when they used any of the two components separately at 0.05%.

Dioxathion was used effectively in Zimbabwe, as shown by Hammant (1977), who emphasizes its importance for the control of ticks of various hosts, resistant to arsenic and toxaphene.

In Argentina, Pizarro et al. (1963) carried out trials with the products used in the tick campaign area, and stressed the efficacy of the drug in different trials treating a high number of head as well as the disturbance of fertility and oviposition of *Boophilus microplus* (Cerineo et al. 1963). However, he found it to be less efficient on engorged nymphs for, although with a delayed evolution, a percentage remained that survived the effect of the product (Cerineo et al. 1963) (Suarez and Serrano 1963). Finally, the author gave evidence that in many establishments, particularly in the provinces of Entre Ríos and Santa Fe where there already was resistance to diazinon, there was also resistance to delnav.

Carbamates

Carbaryl

Commercial Name: Sevin

Laboratory: Union Carbide

Chemical Name: 1-Naphthyl-methylcarbamate

Structural Formula:

$$O-\underset{\underset{O}{\|}}{C}-NH-CH_3$$

Solubility: Practically insoluble in water

Formulation: 45% w/w

Recommended Dosage: 0.2 to 0.3% w/v

Its mode of action is similar to that of the organophosphorous compounds. For this reason it was not totally effective against resistant strains of *Boophilus microplus*, as seen in the field trials carried out by Roulston et al. (1968b) using carbaryl 0.3%.

However, it did have a good effect on susceptible strains, even at half the concentration.

Nolan (1979) confirms this aspect when he points out that these compounds were used successfully in Australia for 4 years.

Unfortunately, as Harrison (1967) points out, this compound could not be developed for large-scale use due to the fact that it progressively loses efficacy in plunge dips subjected to the passage of large numbers of animals (Drummond et al. 1968), and due to the appearance of strains resistant to organophosphorous compounds that also showed a cross-resistance to carbamates.

New Acaricide Drugs

Here we include those drugs that appeared towards the end of the 1970's which have a mechanism of action different from the previous compounds. This mechanism is based on the recognition of strains resistant to the compounds in use.

Although we will describe each one separately, Table 5.4 (taken from Nolan et al. 1977) shows the percentages of efficacy of these new compounds as compared with the former ones.

Nimidane

Commercial Name: Abequito

Laboratory: Cyanamid

Chemical Name: 2-(4-Chloro-o-tolyl)-imine-1,3-dithiethane

Table 5.4. Percentage of control of susceptible strains of *Boophilus microplus*, organophosphorous-resistant and DDT-resistant. Obtained in trials by aspersion with commercial acaricides

Compound	Conc. % w/v	Susceptible	Biarra	Mackay Silkwood	Mt. Alford	DDT-resistant
DDT	0.5	93	93	–	–	–
Coumaphos	0.025	>99	41	69	–	–
Ethion	0.075	>99	74	90	34	–
Chlorpyriphos	0.025	>99	90	99	46	–
Carbaryl	0.2	>99	75	–	–	–
Promacyl	0.15	>99	98	95	99	–
Chlordimephorm	0.1	>99	99	97	99	–
Amitraz	0.025	>99	>99	>99	>99	–
Clenpyrin	0.3	97	99	85	98	–
Chlormethiuron	0.2	97	–	99	97	
Nimidane	0.05	–	–	99	–	
Permethrin	0.025	98	–	–	–	81
Cypermethrin	0.005	93	–	–	–	66
	0.025	98	–	–	–	96
NRDC 161	0.05	>99	–	–	–	85
Deltamethrin	0.025	>99	–	–	–	>99

Structural Formula:

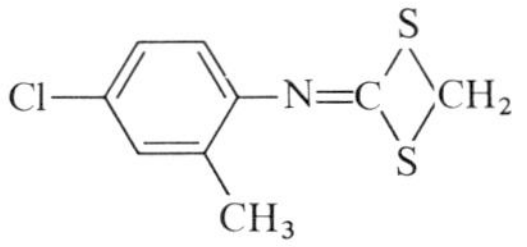

$$Cl-\underset{\underset{CH_3}{|}}{\langle\bigcirc\rangle}-N=C\underset{S}{\overset{S}{\diagdown}}CH_2$$

Chemical Formula: $C_9H_8NS_2Cl$

Chemical Group: Derived from anilines, dithiethane

Formulation: 50% Active drug

Recommended Dosage: Dipwash = 0.10%; replenishment and reinforcement = 0.20%

In 1974 Amaral, N. K., et al. carried out the first trials in Brazil, with nimidane by aspersion at a concentration of 500 ppm. They obtained a 99.6 – 99.8% control of *Boophilus microplus* and almost 96% reduction of oviposition. They also confirmed a residual effect of approximately 15 days.

In later trials this compound also showed a good efficacy against the Biarra strain (Welch 1975).

Without doubt, the best method of application was by aspersion, as shown by the profusion of trials of this type (Angel 1975) and in view of the scarcity of bibliography on its effect by immersion.

Betancourt et al. (1978) observed very good efficacy at 500 ppm on *Boophilus microplus* and *Anocentor nitens*, with a residual effect of 8 days. In Colombia, using concentrations of 500 and 750 ppm, almost 100% control was obtained over *Boophilus microplus*.

Amitraz

Commercial Name: Triatix BTS 27419

Laboratory: Boots

Chemical Name: N,N-bis-(2,4-Dimethylphenylimino-methyl-methylamine)

Chemical Group: Diamidide

Structural Formula:

$$\text{N–CH–N–CH–N} \quad (\text{with CH}_3 \text{ substituents and two dimethylphenyl rings})$$

Chemical Formula: $C_{19}H_{23}N_3$

Solubility: Slightly soluble in water, but very soluble in organic solvents

Formulation: 12.5% w/v of active drug

Recommended Dosage: Dipwash = 0.025%; replenishment and reinforcement = 0.05%

Toxicity: Acute oral LD_{50} in rats = 600 mg kg^{-1}
  Acute oral LD_{50} in rabbits = 1600 mg kg^{-1}.

This compound, used successfully for the treatment of resistant strains of sheep scab (Bridi et al. 1976 – 77) (Muñoz-Cobeñas et al. 1978) (Nuñez 1977), revealed identical conditions for the control of *Boophilus microplus* at 0.025% (Palmer et al. 1971).

Also, different authors worldwide (J. A. F. Baker et al. 1973, Harrison et al. 1973a, Uribe et al. 1976, K. Allan and Palmer 1976) were of the opinion that, added to the direct effective action against resistant tick strains (Wharton and Roulston 1977), was the prolonged residual effect of 14 days and an intensive effect on oviposition.

Thus arose a new alternative for controlling (Harrison et al. 1973b) with an even greater knockdown capacity than chlordimeform (Stone and Knowles 1973).

Its great stability in plunge dips was also described, due to the calcium hydroxide that, generally at 0.5%, is added to the preparation. This compound provides a high pH of 12 that chemically gives stability to the drug, also inhibiting the bacterial proliferation, thus avoiding its degradation that could become more acute due to the presence of *Pseudomonas* spp. and *Achromobacter* spp. (P. B. Baker and Woods 1977) (Allock et al. 1978).

Another of its advantages is its low toxicity to animals: LD_{50} 1600 mg kg^{-1} (Weighton et al. 1973). This permits dipping the very young animals and therefore a general treatment of the population of the farm.

Regarding its mode of action, although not yet fully clarified, it is known that it has an action that opposes monoaminooxidase and also that, after penetrating into the tick, it can metabolize into N-2,4-Dimethylphenyl-N-methylformamidine, which is very toxic, especially to larvae (Schuntner and Thompson 1979).

Clenpyrin

Commercial Name: Bayer 6896

Laboratory: Bayer

Chemical Name: 2-(3,4-Dichlorophenylimine)-1-N-butylpirrolidone

Chemical Group: Cycloamidine

Structural Formula:

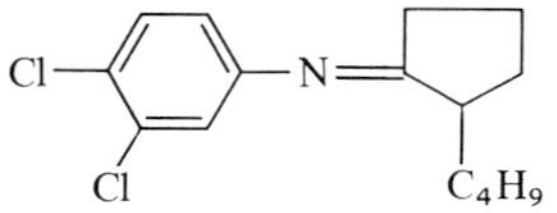

Chemical Formula: $C_{15}H_{19}NCl_2$

Solubility: Almost insoluble

Formulation: 30% w/v of active drug

Recommended Dosage: 0.2 − 0.3% w/v

Toxicity: Acute oral LD_{50} in rats = 2586 mg kg^{-1}
Acute oral LD_{50} in calves = 250 mg kg^{-1}
Acute dermic LD_{50} in cattle = 20,000 ppm.

This new compound proved to be active against strains resistant to organophosphorous compounds, in trials in vitro by infestation and by contact.

In their work, Enders et al. (1973) describe a 100% inhibition of oviposition and the paralysing effect 24 h after treatment. Enders is of the opinion that the mode of action is not due to the inhibition of cholinesterase, but that the lethal effect is caused by the complete paralysis of the muscles, making it impossible for the tick to feed. On the other hand, it inhibits the development of the ovary, thus preventing the formation of eggs.

Nolan (1979) also stresses the stability of the compound in contaminated plunge dips, as well as a wide safety margin for the animals treated. In his paper he mentions the observations made by Andrews and Stendel, that the muscular paralysis is due to the blocking of all the processes of synthesis by means of the inhibition of NADH oxidization.

Stendel and Andrews (1973) carried out trials in vitro at concentrations of 1500 ppm, and obtained 100% efficacy. Nevertheless, to control the Mackay strain in vitro, between 2000 and 2500 ppm were necessary, and even so a few engorged nymphs survived treatment.

Chlordimeform

Commercial Name: Chlorfenamidine

Laboratory: Ciba-Geigy

Chemical Name: N-(2-Methyl-4-chlorophenyl)-N'N'-dimethyl-formamidine

Chemical Group: Formamidine

Structural Formula:

$$Cl-\underset{}{\bigcirc}\overset{CH_3}{-}N{=}CH{-}N(CH_3)_2$$

Chemical Formula: $C_{10}H_{13}N_2Cl$

Solubility: Soluble in organic solvents

Formulation: 50% w/v of active drug

Recommended Dosage: 0.025 – 0.1%

Toxicity: Acute oral LD_{50} in rats = 250 mg kg^{-1}
Acute dermic LD_{50} in rats = 4000 mg kg^{-1}
LD_{50} in cattle = unknown.

According to trials in vitro carried out by Gothe and Hartig (1976), this compound proved to be very effective against strains such as Biarra, Mt. Alford, Mackay and those resistant to carbamates, delaying oviposition of engorged females and embryo development.

However, Knowles et al. (1973) found a certain tolerance in larvae of the carbamate strain.

Regarding its mode of action, Nolan (1979) indicates that the symptoms of intoxication with chlorfenamidine in ticks are of sympathetic-mimetic nature, and due to monoaminooxidase interference.

Knowles and Roulston (1972) also supported this, after observing the ticks treated. Atkinsons and Knowles (1974) found high levels of monoaminooxidase in homogenates of ticks, demonstrating the presence of this enzyme in salivary glands, peripheric nerves and ganglions.

Roulston et al. (1971) describe the behaviour of the adult forms of the parasite cycle, saying that they become detached from where they were fixed, move along the body surface and then fall and die within a short time. The larvae affected by the compound become immobilized, they move their limbs uncoordinatedly, go into coma and take quite a long time to die. This fact induced Knowles et al. (1973) to modify his techniques for studying the sensibility of the compound to larvae of the Mackay strain. He prolonged the period of contact of larvae with paper impregnated with the drug up to 144 h, thus confirming the low level of resistance.

However, despite the good results shown by other trials, this compound was not developed commercially, due to its relative toxicity to cattle and its instability in plunge dips.

For this reason, mixtures with other compounds were studied. By adding only 0.01% chlorofenamidine to organophosphorous and arsenical acaricides, Roulston et al. (1971) obtained better results over the Biarra strain, in comparison with the effect of these active ingredients separately.

Chlormethiuron

Commercial Name: C_{9140}-Dipofeno

Laboratory: Ciba-Geigy

Chemical Name: 1-(4-Chloro-2-methylphenyl)-3,3-dimethylthiourea

Chemical Group: Thiourea

Structural Formula:

$$Cl-\underset{\underset{CH_3}{|}}{C_6H_3}-NH-\underset{\underset{S}{||}}{C}-N\underset{CH_3}{\overset{CH_3}{<}}$$

Chemical Formula: $C_{10}H_{13}N_2SCl$

Solubility: Very soluble in acetone; insoluble in water

Formulation: 60% w/v active drug

Recommended Dosage: 0.18% w/v

Toxicity: Acute oral LD_{50} in rats = 2500 mg kg^{-1}
Acute dermic LD_{50} in rats = 2150 mg kg^{-1}
LD_{50} in cattle = not known.

This derivative of thiourea has a cross-resistance with chlordimeform, for which reason it is suspected of having a similar mechanism of action.

In this respect Schuntner and Thompson (1979) carried out toxicity and metabolism studies of the compound on *Boophilus microplus* larvae resistant to organophosphorous compounds. In Queensland, where it has been used for 7 years, they proved its excellent efficacy and the possibility of prolonging its use in the field.

Von Orelly et al. (1978) point out its stability and low toxicity, that make possible the use of high concentrations, as well as the cleansing of cattle at 0.2% and a good inhibition of oviposition.

Iminothiazol

Commercial Name: Tifatol

Laboratory: Ciba-Geigy

Chemical Name: 2-(2′,4′-Dimethyl-phenylimine)-3-methyl-4-thiazoline

Chemical Group: Iminothiazol

Structural Formula:

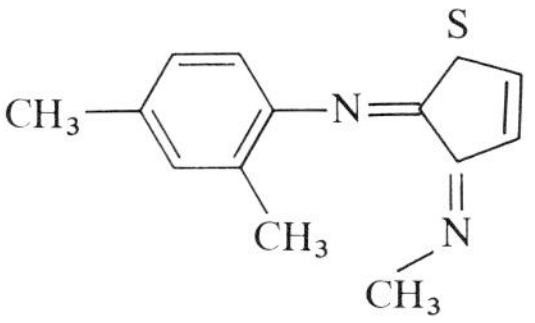

$$CH_3 - \text{[2,4-dimethylphenyl]} - N = \text{[thiazoline ring: S, N-CH}_3\text{]}$$

Chemical Formula: $C_{14}H_{14}N_2S$

Solubility: Good with organic solvents, less in water

Formulation: 30% w/v active drug

Recommended Dosage: Dipwash = 0.03%; replenishment and reinforcement = 0.09%

Toxicity: Acute oral LD_{50} for rats = 725 mg kg^{-1}
Acute percutaneous LD_{50} for rats = 3100 mg kg^{-1}
Oral LD_{50} in cattle = unknown.

This compound showed good results in the preliminary trials with *Boophilus microplus* strains resistant to organochloride and organophosphorus compounds.

These results were confirmed in Argentina with naturally infested animals, also evaluating the inhibition caused to oviposition of engorged females which, although not complete, inhibits eclosion of eggs.

These trials helped resolve the need to carry out reinforcements of 150% in order to improve the given results.

Regarding the residual effect shown, the bibliography consulted mentions a period of 72 h during which larvae reinfestation is avoided.

Synthetic Pyrethroids

In ancient times, the Persians already knew of the insecticidal action of natural pyrethrum and originally prepared it by grinding *Chrysanthemum cinerariaefolium* flowers, obtaining as time passed preparations with a larger concentration of the active ingredients.

Their characteristics rapidly became widely known and pyrethrum was grown in areas where the climate was adequate, such as the Balkans, Kenya and Ecuador. However, due to the high cost of picking the flowers, its use was restricted for a long time to the domestic area.

In 1935, the insecticide efficacy of synthetic pyrethroids was described as well as their low toxicity to mammals, but they were still not developed due to their instability in light, their high cost, and partly due to the success of the more economical synthesis drugs such as organochlorides and carbamates.

Nevertheless, the resistance that developed to these latter compounds, as well as the harmful residues they left, motivated new investigations and their wide introduction in veterinary medicine.

Only in 1965, with the aid of gaseous chromatography, did the chemical structures of the six compounds with insecticide activity contained in pyrethrum became known. These are Pyrethrin I and II, Cinerin I and II and Jasmolin I and II.

	R_1	R^2
Pyrethrin I	$-CH_3$	
Pyrethrin II	$-COOCH_3$	
Cinerin I	$-CH_3$	
Cinerin II	$-COOCH_3$	
Jasmolin I	$-CH_3$	
Jasmolin II	$-COOCH_3$	

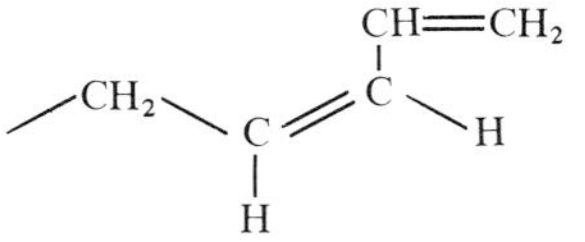

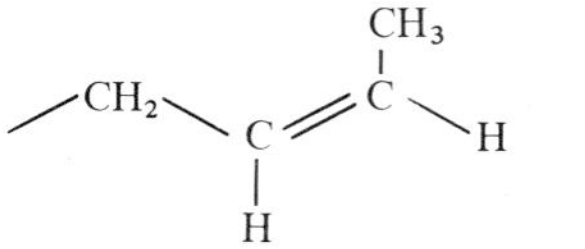

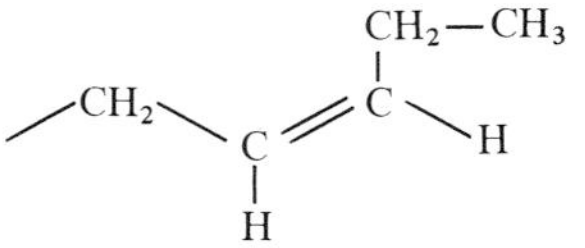

All these compounds are esters of the chrysanthemic acid or the pyrethric acid (chrisanthen dicarboxilic) and alcohols (retrolones).

Likewise, the structure of the chrisanthemic acid has the possibility of 4 configurational and geometrial isomers commonly known as d-cis, 1-cis, d-trans and

1-trans. The pyrethric acid has the possibility of geometric isomery with the alkenic remains, giving rise to a total of 8 estherisomers. Similarly, alcohols can generate 2 configurational isomers and a pair of geometric isomers; Cinerin II has 32 possible isomers with different insecticide activity.

Evolution of the Synthetic Pyrethroids

In 1949 Schechter et al., by modifying the lateral chain of cetonic alcohol, synthesized the first pyrethroid of practical importance and named it allethrine. From there a mixture of 8 optically active isomers was obtained, the most effective being d-allethrine, bioallethrine and s-allethrine, which are characterized by their capacity for knockdown rather than for their lethal effect on insects.

Allethrine structure

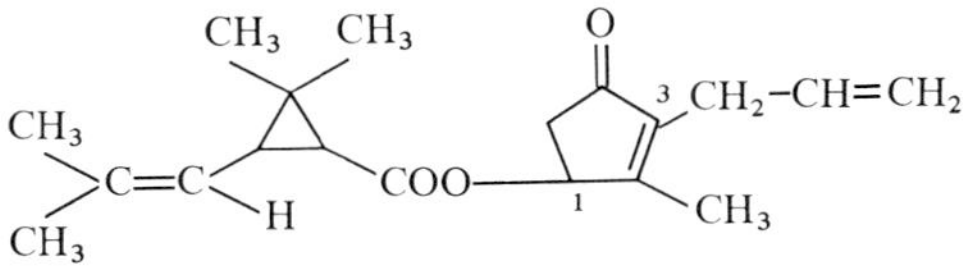

In 1964 Kato selected tetramethrin from a series of esthers of the chrysanthemic acid for its efficacy. However, it was not developed because is also proved to act fundamentally as a knockdown agent.

Tetramethrin

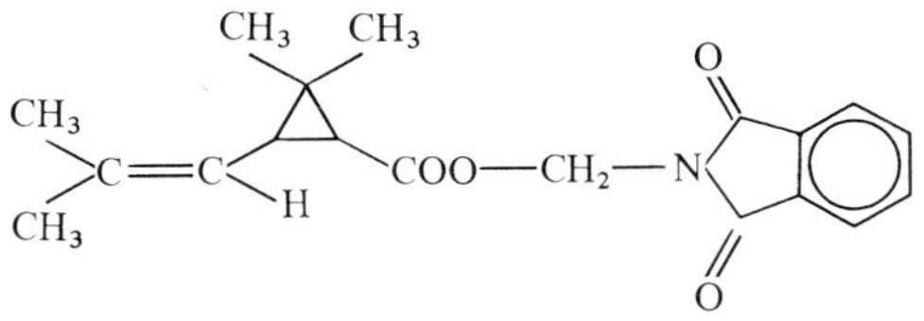

Looking for a compound with more insecticidal effect, in 1967 Elliot found it in the esther of the alcohol 5-benzyl-3 furyl-methylic, or biorresmethrin, but is was not further developed, mainly due to its loss of efficacy on warm-blooded animals.

Biorresmethrin

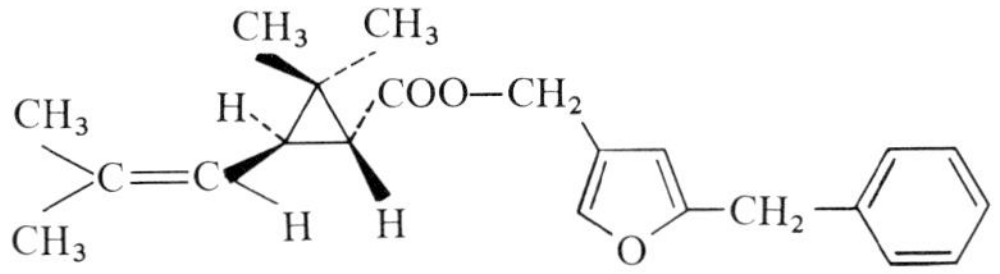

With the incentive of the search for new alcohols, in 1973 Elliot et al. discovered 3-phenoxibenzylic, known as Elliot Alcohol, which was later used by Fujimoto in the synthesis of phenothrin, of great photostability.

Elliot Alcohol

He finally obtained permethrin by modifying chrysanthemic acid for which he replaced the methyl groups attached to the vinyl side chain by halogen atoms, followed by estherification with a z-phenoxybenzyl alcohol.

Permethrin

With this compound, at present widely used, it is possible to combine an insecticide effect, a low toxicity to mammals and a fundamental persistence due to its excellent stability to light and air.

Elliot and his collaborators also obtained the synthesis of cypermethrin by introducing a cyano group on the carbon present in the alcoholic hydroxide (cyanhidrine) in an S configuration.

Cypermethrin

A year later, in 1974, Ohno obtained phenvalerate by using the same alcoholic fraction but altering the structure of the acid. Although the phenvalerate does not have the cyclopropylic remains, it maintains a high insecticide effect.

Phenvalerate

That same year, Elliot obtained the synthesis of deltamethrin by changing the structure of cypermethrin.

Insecticide Effect

The abundant bibliography of different authors describes the excellent insecticide effect of these compounds (Martin 1975, Matter et al. 1976, Schmidt et al. 1976, Blackman and Hodson 1977, Elliot et al. 1978).

These insecticides are very versatile, some of low and some of very high stability, with long or short residual effect, with a rapid or slow lethal effect, but all of them of low toxicity and rapid metabolization in domestic animals.

Tickicide Effect

The first studies carried out by Larkin in 1958 (see Larkin 1961), using a simple aqueous formulation of pyrethrins at 0.01% and 0.0125% against *Boophilus microplus*, proved the versatility and quality of the pyrethroids. In a similar way, Schuntner et al. (1974), using pyperonyl butoxide alone or mixed with carbaryl, obtained a 99% control over the Biarra strain of *Boophilus microplus*.

In a personal communication, M. Matthewson (Wellcome Research Laboratories, Berkhamsted, England) describes the high activity of permethrin in laboratory and field trials against different tick species such as *Boophilus microplus, Boophilus decoloratus, Rhipicephalus appendiculatus, Rhipicephalus eversti* and *Ammblyoma hebraeum*.

On strains resistant to organophosphorous tickicides and carbamates, we must mention studies carried out by Nolan et al. (1977) (1979) that show the excellent activity against *Boophilus microplus* and emphasize deltamethrin as one of the most active compounds of the group, even showing an acceptable effect against strains resistant to DDT.

This last compound was used successfully by Bulman et al. (1980) in Formosa, Argentina, against *Boophilus microplus* in plunge dips where a large number of head of cattle were treated during 13 months.

The most recent synthetic pyrethroid was evaluated by Stubbs et al. (1982) in the control of the major resistant strains of the cattle tick (*Boophilus microplus*) and for control of the Buffalo Fly (*Haematobia irritans exigua*) on cattle. At an 0.007% concentration of cyhalothrin with regulated treatments, they obtained over 99% control of Biarra, Mackay, M. T. Alford and DDT-resistant strains. They also developed a study of the period of residual protection, spraying animals with 0.007% cyhalothrin, which then received a heavy challenge from the Biarra tick strain. These studies showed that cyhalothrin provides a residual protection against tick reinfestation of 7 to 15 days, and of more than 28 days against buffalo fly reinfestation.

This information was reconfirmed by field trials treating animals every 28 days, observing that the reduction of tick levels in pastures makes it possible to lengthen the interval between treatment up to 7 weeks.

Mode of Action

Although the mode of action of natural and synthetic pyrethroids is not fully defined, it is known that they are never toxic and that they do not act by interfering with the acetylcholine.

Nevertheless, due to their physical properties (Brigs et al. 1974), it is considered that they act on the nerve cell membrane, causing changes in the permeability of the sodium and potassium ions and producing an increase of the negative potential, with repetition of discharges, followed by hyperexcitation and posterior blockage of the nervous conductor, which leads to the paralysis of the intoxicated ixodide.

Toxicity to Man and Animals

From the studies carried out (Leahey 1979) it has been determined that these compounds are degraded by the enzymatic systems of mammals and by the microorganisms present in the soil, for which reason they do not contaminate the atmosphere nor do they leave residues in the animals treated that might be harmful to man.

Below we transcribe the structural formulae of the most recent pyrethroids.

Permethrin

Chemical Name: 3-Phenoxybenzyl ($\pm$) cis/trans-2,2-dimethyl-3-(2,2-dichlorovinyl)cyclopropane-1-carboxylate

Structural Formula:

Chemical Formula: $C_{20}H_{20}O_3Cl_2$

Recommended Dosage: 0.025 and 0.10%

Toxicity: Acute oral LD_{50} for rats = 3185 mg kg^{-1}
      Acute oral LD_{50} for rabbits = 5500 mg kg^{-1}
      Abomasum injection in ruminants = 400 to 800 mg kg^{-1}.

Cypermethrin

Chemical Name: $(\pm)$-α-Cyano-3-phenoxybenzyl-$(\pm$ cis-trans-2,2-dimethyl-3-(2,2-dichlorovinyl)cyclopropane-1-carboxylate

Structural Formula:

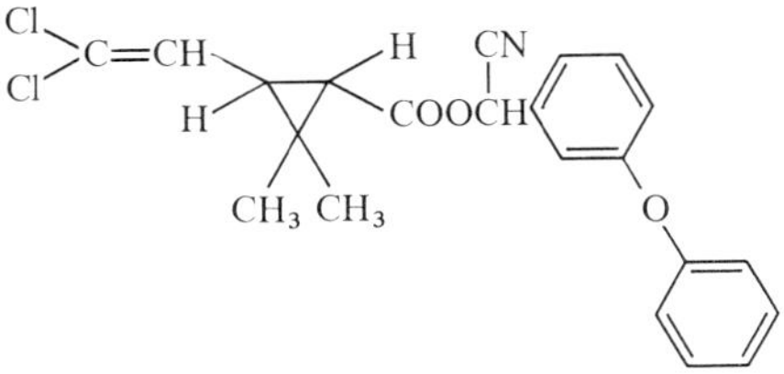

Chemical Formula: $C_{22}H_{19}O_3NCl_2$

Recommended Dosage: Dipwash = 0.15%; replenishment and reinforcement = 0.25%

Toxicity: Acute oral LD_{50} for rats = 251 mg kg^{-1}.

Cypothrin

Chemical Name: Espiro(cyclopropane-1-1′indeno)-2-carboxylicacid,3,3-dimethyl-α-cyano-m-phenoxybenzyl esther

Structural Formula:

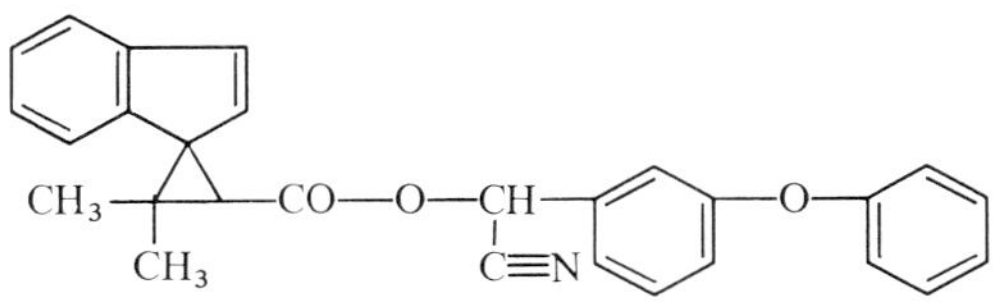

Chemical Formula: $C_{28}H_{23}O_3N$

Recommended Dosage: Dipwash = 0.17%; replenishment and reinforcement = 0.24%

Toxicity: Acute oral LD_{50} for rats = 1203 mg kg^{-1}
      Acute percutaneous LD_{50} for rabbits > 5000 mg kg^{-1}
      Acute oral LD_{50} for cattle = 100 mg kg^{-1}.

Cyhalothrin

Chemical Name: 2,2-Dimethyl-3-(2-chloro-3,3,3-trifluoro-1-propenyl)-cyclopropane carboxylate of α-cyano-3-phenoxybenzyl

Structural Formula:

$$CF_3ClC=CH—\overset{\overset{\displaystyle H}{|}}{\underset{\underset{\displaystyle CH_3}{|}}{C}}—\overset{\overset{\displaystyle H}{|}}{\underset{\underset{\displaystyle CH_3}{|}}{C}}—COO\overset{\overset{\displaystyle CN}{|}}{\underset{\underset{\displaystyle H}{|}}{C}}$$

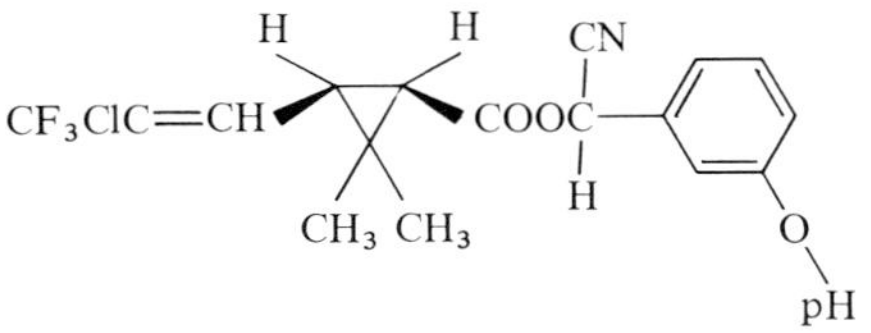

Chemical Formula: $C_{23}H_{19}NO_3ClF_3$

Recommended Dosage: Dipwash = 0.005%; replenishment and reinforcement = 0.010%

Flumethrin

Chemical Name: 2,2-Dimethyl-3-(2-(4-chlorophenyl)2-chlorovinyl)-cyclopropylcarboxylate of α-cyano-4-fluoro-3-phenoxybenzyl

Structural Formula:

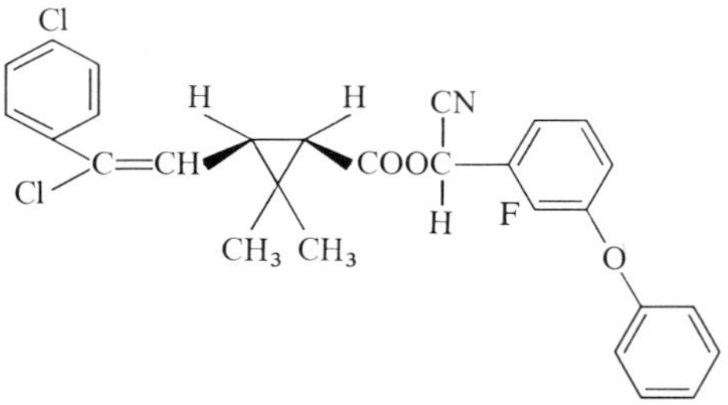

Chemical Formula: $C_{28}H_{22}Cl_2FNO_3$

Recommended Dosage: Dipwash = 0.0030%; replenishment and reinforcement = 0.0030%

Toxicity: Acute oral LD_{50} for rats = 10,000 mg kg^{-1}
Acute oral LD_{50} for rabbits = 10,000 mg kg^{-1}
Acute intramuscular LD_{50} for rats = >2000 mg kg^{-1}.

Deltamethrin

Chemical Name: 2,2-Dimethyl-3-(2,2-dibromovinyl)-cyclopropyl-carboxylate of α-cyano-m-phenoxybenzyl

Structural Formula:

Chemical Formula: $C_{22}H_{19}O_3NBr_2$

Recommended Dosage: Dipwash = 0.0030%; replenishment and reinforcement = 0.045%

Toxicity: Acute oral LD_{50} for rats = 139 mg kg^{-1}
  Acute percutaneous LD_{50} for rabbits = 2000 mg kg^{-1}.

This new range of acaricides opens broad perspectives that may be of great influence in the fight for the control of *Boophilus microplus,* only if very stable formulations are obtained that successfully resist the tough conditions of its use in the fields.

Drugs of Systemic Action

Ivermectin

Commercial Name: Ivomec

Laboratory: Merck Sharp & Dhome

Chemical Name: 22,23-Dehydroavermectin B_1

Structural Formula:

Chemical Formula: 22,23-dihydroavermectin B_1A $C_{48}H_{74}O_{14}$ (not more than 80%); Molecular weight: 845.10. 22,23-dihydroavermectin B_1B $C_{47}H_{72}O_{14}$ (not less than 20%); Molecular weight 861.07

Formulation: Injectable solution 1% ivermectin

Recommended Dosage: 200 µg kg^{-1} body weight. It must be injected subcutaneously at a rate of 1 mg/50 kg w/v

Toxicity: Acute oral LD_{50} for rats = 87.2 mg kg^{-1}
Acute percutaneous LD_{50} for rabbits = 406 mg kg^{-1}.

The avermectins are derived from the fermentation of a soil microorganism denominated *Streptomyces avermetilis*. Certain derivatives of avermectin were studied by scientists of MSD, choosing ivermectin for development.

This novel drug is characterized by its broad spectrum action on endo and ectoparasites that feed on the blood stream. The mode of action on arthropods is based on the inhibition of transmission of signals in the neuromuscular joints, stimulating the discharge of gamma-amine butyric acid (GABA), neurotransmitter, inhibitor of the presynaptic nervous terminals.

In Argentina, Lombardero and his team (1984) carried out trials to confirm the effect of the drug. They used cows naturally infested with *Boophilus microplus* to which they made only one application at 1%.

The daily examinations carried out during 3 consecutive weeks on the treated animals and the controls provided significant results, fundamentally in the percentage of decrease in size of the engorged females, oviposition and hatching. The figures obtained are the following:

	Treated lot	*Control lot*	*Reduction (%)*
Size of engorged females	4.8 mm	8.9 mm	46.0
No. of eggs	133	1948	93.2
Larvae hatching (%)	32.3	97.5	66.9

The most relevant findings were that as from the third and fourth days posttreatment there were no fully developed females, the oviposition being proportional to the size of ticks from the treated group. Although reduction started from the first day of the trial, even disappearing for a few days, it reappeared towards the end of the trial. Regarding the hatching of larvae, although their number was markedly reduced, they were never totally annulled, the lowest figure being 1.7% on the 9th day of the trial (see Figs. 5.13 and 5.14).

These authors also studied the acaricide effect of the drug on different stages of *Boophilus microplus*, treating a group of naturally and artificially infested cattle on three occasions and with 5-day intervals between inoculations. They

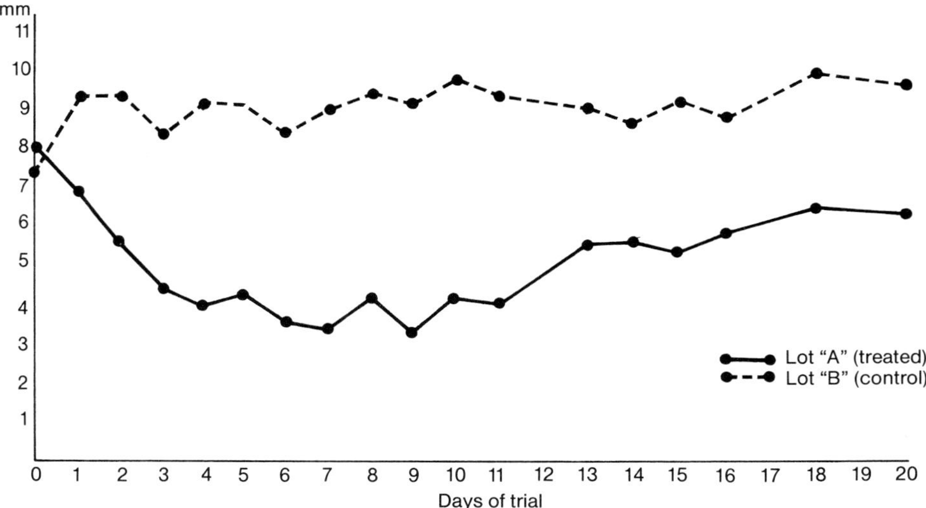

Fig. 5.13. *Boophilus microplus*, size of females, daily averages

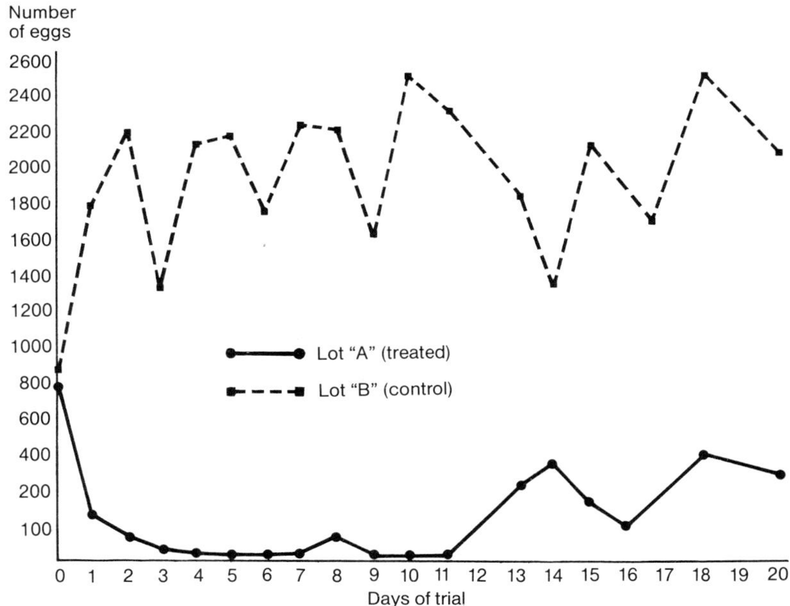

Fig. 5.14. *Boophilus microplus*, number of eggs per female

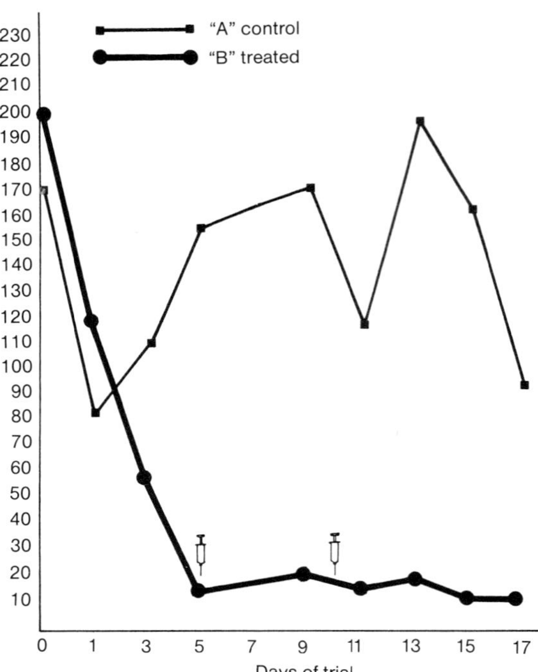

Fig. 5.15. Average count of all *Boophilus microplus* stages per lot

found an efficacy variable between the following figures: on females of 4.5 – 8 mm, between 93.81% and 100%; on nymphs and engorged females 100% efficacy on days 5 and 3 of the trial respectively. While on larvae the activity was low or irregular with barely a maximum efficacy of 36.46% (see Fig. 5.15).

Nolan et al. confirmed that a single dose of 200 μg kg^{-1} w/v applied subcutaneously to naturally infested cattle provides a satisfactory tick control for 3 weeks.

However, they pointed out that during the first day after application, a significant number of engorged females survived the treatment without the viability of oviposition being affected. This phenomenon could be undesirable for two reasons: first, when the cattle must be taken to pasture or areas free of ticks immediately after treatment, and second, because this survival could increase the level of selection of resistant strains.

In another part of the trial, the authors carried out an evaluation on the compound, in small doses applied daily subcutaneously, finding the 15 μg kg^{-1} per day provided an excellent tick control. For this reason they suggested the possibility of using the drug with a subcutaneous implant of slow release, which could release small quantities continuously over a determined period of time. For example, 1 g of active ingredient releasing 15 μg kg^{-1} per day in a 300-kg animal would provide protection for 200 days. This could permit strategic treatments, covering the periods with the heaviest tick load, such as summer and spring.

Finally Nolan (1981) proposed that to transfer cattle to an area free of ticks, satisfying the official requirements of a complete cleansing of the animals to be transferred, it would be necessary to apply two subcutaneous injections of 200 µg kg^{-1} with an interval of 3 days between applications.

However, following observations made by Lombardero (1984) and previously mentioned, two applications would not be sufficient to free the animals from all stages of the life cycle.

According to the regulations in force in Argentina for the aggressive development of the sanitary campaign to control tick infestations, this schedule of treatments would not surpass the high demands imposed for the movement of cattle to tick-free areas.

Closantel

Commercial Name: Seponver

Laboratory: Johnson & Johnson − Janssen Pharmaceutica M.V.

Chemical Name: N-{S-Chloro-4-[(4-chlorophenyl)-d-cyanomethyl]-2-methyl-phenyl}-2-hydroxi-3,5-diiodo-benzamide

Structural Formula:

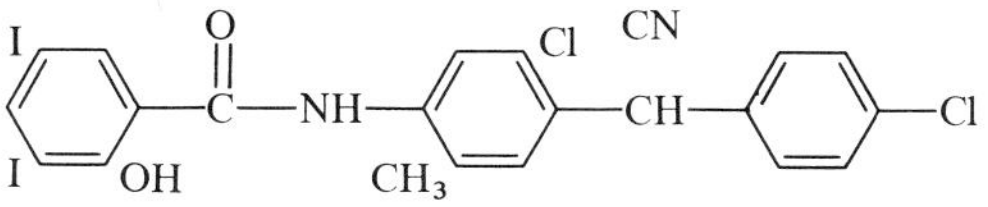

Chemical Formula: $C_{22}H_{14}Cl_2I_2N_2O_2$

Formulation: Injectable solution at 30%

Recommended Dosage: 22.5 mg kg^{-1} body weight. It must be administered orally at a rate of 1 ml per 13.3 kg

Toxicity: LD_{50} for ruminants = 40 mg kg^{-1} w/v.

Closantel was developed by the laboratory of investigation of Janssen Pharmaceutica of Belgium. Its main characteristic is that of being active when applied orally or by injection against some endo and ectoparasites that feed on blood of lymph.

Its mechanism of action is to alter the oxidative phosphorylation, thus inhibiting the transport of electrons through the respiratory chain.

As a tickicide Janssen recommends repeating the dose every 30 days. However, according to Bulman et al. (1981) to obtain a total cleansing of cattle naturally infested with *Boophilus microplus*, in their trials they had to inoculate the animals four times with closantel 5%, with doses varying between 10 and 15 mg kg^{-1}, for 1 month. The authors also confirmed that the drug remains active between 2 and 3 weeks, practically disappearing between the 3rd and 4th weeks.

Another study they carried out was the acaricide effect on engorged females of cattle inoculated at intervals of approximately 30 days, in a schedule of five inoculations, observing that up to 12 days post-inoculation only 80% of the engorged females died before completing their normal cycle. The inhibiting action on larvae hatching reached 64.22%, that means it was not complete. These data are in accordance with those described by Lombardero et al. (1980) who inoculated cattle at dosis of 20 mg kg^{-1} in order to evaluate the systemic effect of closantel injectable 5%, over oviposition and hatching of a *Boophilus microplus* strain. They pointed out that the compound did not affect oviposition, but did reduce the percentage of hatching, and that, although with one dose there was a decrease of elements that reached the engorged female stage, there was no lethal effect on adult ticks.

The efficacy of the drug is greater against larvae and nymphs, while in the evolutional stages its activity is less and with one application parasite elimination is never complete.

References

Allan J (1955) Loss of biological efficiency of cattle dipping. Nature (London) 175:1131 – 1133

Allan K, Palmer W (1976) The ixodicidal efficiency of a number of pour-on formulations of Amitraz against the Biarra strain of the southern cattle *Boophilus microplus* on housed calves. Proc Int Conf Tick Borne Dis Vectors, Edinburgh, 27 Sep – 1 Oct 1976, pp 241 – 18

Allock ER, Woods DR, Rivet DEA (1978) Bacterial degradation products of the ixodicide Amitraz. J Appl Bacteriol 44:383 – 386

Amaral NK, Monmany LFS, Carvalho LAF (1974) Acaricide A.C. 84633: First trials for control of *Boophilus microplus*. J Econ Entomol 67 (3):387 – 389

Angel ECA (1975) Efficacy of Abequito* Nimidane and Cygon* dimethoate against ticks. American Cyanamid Co. International Agric Res and Dev Rep Jan – February:1 – 4

Armstrong G, Bradbury FR, Standen H (1951) The penetration of the insect cuticle by isomers of benzene hexachloride. Ann Appl Biol 38:555 – 556

Atkinsons PW, Knowles CO (1974) Induction of hyperactivity in larvae of the cattle tick *Boophilus microplus* by formamidines and related compounds. Pestic Biochem Physiol 4:417 – 424

Ault CN (1946) Control of *Boophilus australis* in the Argentine by the gamma isomer of Hexachloro cyclohexane (Gammexane). Nature (London) 157:699 – 700

Ault CN (1948) Investigaciones sobre las dificultades de combatir la garrapata *Boophilus microplus*. Rev Med Vet 30:1 – 52

Ault CN (1950) El empleo del canfeno clorado (Toxafeno) contra la garrapata *Boophilus microplus Can*. Gac Vet XII Buenos Aires 67:208 – 213

Australia Res (1919) The cattle tick pest. Inst Sci Ind Bull 13:15 – 19

Azis SA, Knowles CO (1973) Inhibition of Monoamine oxidase by the pesticide chlordimeform and related compounds. Nature (London) 242:417 – 418

Baker JAF, Thompson GE (1966) Supona (chlorfenvimphos) for cattle tick control. Part I. Hand spraying trials. J S Afr Vet Med Assoc 3:367 – 372

Baker JAF, Shaw RD, Thompson GE, Thoburn JA (1962) Development of a composite dioxathion/chlofenvinphos concentrate for control of single and multi-host ticks and their resistant strains in Southern Africa. Soc Chem Ind Monogr 33:172 – 181

Baker JAF, Taylor RJ, Stanford GD (1973) A new triazapenta diene compound active against cattle ticks of mayor importance in South Africa. Proc 7th Br Insectic Fungicide. Conf, pp 291 – 300

Baker PB, Woods DR (1977) Cometabolism of the ixodicidal amitraz. J Appl Bacteriol 42:187 – 196

Barnett SF (1961) The control of ticks on livestock. FAO Agric Stud 54:1 – 113

Barlow F, Elliot M, Farnham AW, Hadaway AB, Janes NJ, Needham PH, Wickham JC (1971) Insecticidal activity of the pyrethrins and related compounds. IV Essential features for insecticidal activity in chrysanthemates and related cyclopropane esters. Pestic Sci 2:115 – 118

Bekker PM (1960) History of cattle dipping. Veld (Spring Issue) part 1:1 – 3, part 2:11 – 14, part 3:9 – 13, part 4:3 – 6

Betancourt AE, Parra D, Angel CA (1978) Evaluación de la efectividad del acaricida Nimidane en el control de *Boophilus microplus*. Rev ICA Bogotá Colombia 2:357 – 361

Blackman GG, Hodson MJ (1977) Further evaluation of permethrin fly control. Pestic Sci 8:270 – 273

Blakeslee EB, Bruce WG (1948) DDT to control cattle tick. J Econ Entomol 41:104 – 105

Blin RC, Gunther FA (1955) Determination of residues of 0,0-diethyl-0-(2-isopropyl-6-methyl-4-pyrimidil) phosphorothidate in milk. J Agric Food Chem 3:1013 – 1016

Boero JJ (1969) Terapéutica antiparasitaria. Ed Univ Buenos Aires, pp 69 – 87

Bridi AA, Bonetti AL, de Souza LAM, Uribe LF (1976 – 1977)) Control of sheep scab outbreak suppusedey resistant to organochlorines and organophosphorus compounds using Amitraz 12.5% E.E. Arq Fac Vet UFRGS/Porto Alegre 4 – 5:135 – 145

Brigs GG, Elliot M, Farnham AW, Janes NF (1974) Structural aspects of the knockdown of pyrethoids. Pestic Sci 5:643 – 649

Brooks FA (1947) Agricultural engineering. Am Soc Agric Eng St Joseph Mich 28:233 – 237

Brown AWA (1951) Insect control by chemicals. Wiley, New York, pp 45 – 67

Brundrett HM, Richards R, Smith CL (1955) Test of toxaphene, chlordane and strabane against lone star tick of cattle. J Econ Entomol 48:233 – 234

Bulman GM, Aguilar M, Diaz CR, Brunel CM, Cicuta ME, Etchechoury MM (1980) Evaluación de la acción garrapaticida de un nuevo piretroide sintético fotoestable en un rodeo de zona infestada de *Boophilus microplus Can*, en el área subtropical de la República Argentina. Gac Vet Buenos Aires 42:338 – 351

Bulman GM, Schmied LM, Aloisi GE, Diaz CR, Brunel CM, Cicuta ME, Etchechoury MM (1981) Resultados de la acción garrapaticida del closantel en solución al 5% inyectable ante la garrapata común del vacuno *Boophilus microplus* (Can) en bovinos de la zona subtropical argentina, conforme a dos esquemas distintos de tratamientos. Gac Vet Buenos Aires 43:359

Cerineo JM, Bocalandro AG, Pizarro IC (1963) Nuestros puntos de vista a las consideraciones formuladas a un trabajo sobre los "Garrapaticidas Organofosforados". Gac Vet Buenos Aires 160:466 – 477

Claborn HV (1956) Malathion in milk and fat from sprayed cattle. J Agric Food Chem 4:941 – 942

Clark AG, Hitchcock M, Smith JN (1956) Metabolism of gammaxane in flies, ticks and locusts. Nature (London) 1:103 – 104

Coollier BL, Dieter CE (1965) Dursban, a new insecticide for chinch bug and sod webworm control in St. Augustine Grass. Down Earth 21 (3):3 – 9

Cooper FA (1962) Delnav (2:3-p-Dioxane-S-bis-(0,0, diethyl dithiophosphate) as an ixodi-
   cide. Vet Rec 4:103 – 113
Csiro (1967) New acaricides control in the Biarra cattle tick. Rural Res CSIRO 60 Sept
   1 – 5
Danilo Parra G (1969) Efecto comparativo del Dursban y del Asuntol. Rev Inst Colomb
   Agropecu 2:73 – 80
Davie MS, Chadwick PR, Hoborn JM, Stewart DC, Wickham JC (1970) Effectiveness of
   the (+)-trans-chrysanthemic acid ester of (+)-allethrolone (bioallethrin) against four
   insect species. Pestic Sci 1:225 – 227
Downing V, Harbour HE, Stones LC (1952) Modern insecticides and ectoparasite control.
   Vet Rec 64:781 – 804
Dresden O, Krijgsman BJ (1948) Experiments on the physiological action of contact insec-
   ticides. Bull Entomol Res 38:575 – 578
Drummond RO (1958) Laboratory screening test of animals systemic insecticides. J Econ
   Entomol 51 (4):425 – 427
Drummond RO (1966) Further evaluation of animal systemic insecticides. Entomol Res
   Agric Res Serv USDA. Kerrvice Tex 1:7 – 12
Drummond RO, Moore B, Warren J (1959) Test with insecticides for the control of the
   winter tick. J Econ Entomol 52 (6):1220 – 1221
Drummond RO, Moore B, Wrich JJ (1960) Field tests with insecticides for control of lone
   star ticks in cattle. J Econ Entomol 53 (5):953 – 955
Drummond RO, Graham OH, Meleney WP, Diamant G (1964) Field test in Mexico with
   new insecticide and arsenic for the control of *Boophilus* ticks on cattle. J Econ Entomol
   57 (3):340 – 346
Drummond RO, Whetstone TM, Ernst SE (1966) Control of ticks on cattle with toxaphene applied by power sprayer and spray race. J Econ Entomol 59 (2):471 – 482
Drummond RO, Ernst SE, Trevino JL, Graham OH (1967) Insecticides for control of the
   cattle tick and the southern cattle tick on cattle. J Econ Entomol 61 (2):467 – 469
Drummond RO, Ernst SE, Trevino JL, Graham OH (1968) Insecticides for control of the
   cattle tick and the southern cattle tick on cattle. J Econ Entomol 61:934 – 936
Drummond RO, Miller JA, Whetstone TM (1981) Control of ticks systematically with
   Merck MK-933, and avermectin. J Econ Entomol 74:432 – 436
Dutoit R, Graf H, Bekker PM (1941) Resistance to arsenical as displayed by the single host
   blue tick, *Boophilus decoloratus* (Koch) in a localized area of the Union of South Afri-
   ca. J S Afr Vet Med Assoc 12 (2):50 – 81
Elliot M, Farnham AW, Janes NF, Needham PH, Pulman DA, Stevenson JH (1973a)
   NRDC 143, a more stable pyrethroid. Proc 7th Br Insectic Fungic Conf 7:7211 – 7228
Elliot M, Farmham AW, Janes NF, Needham PH, Pearson BC (1967) 5-Benzyl-3-3-furyl-
   methyl chrysanthemate: a new potent insecticide. Nature (London) 213:493 – 494
Elliot M, Farnham AW, Janes NF, Needham PH, Pulman DA, Stenvenson JH (1973b) A
   photostable pyrethroid. Nature (London) 246:169 – 170
Elliot M, Janes NF, Potter C (1978) The future of pyrethroids in insect control. Annu Rev
   Entomol 23:443 – 469
Enders E, Stendel W, Wollweber H (1973) New compounds active against resistant cattle
   ticks (*Boophilus* spp.): Relationship between structure and activity within the group of
   cyclic amidines. Pestic Sci 4:823 – 838
Fiedler OGH, Veldman FJ (1957) Asuntol, a new insecticidal compound capable of con-
   trolling all South Africa cattle ticks. J S Afr Vet Med Assoc 28 (3):249 – 253
Fiedler OGH, Vuuren PJJ (1966) Bromophos-ethyl, a new compound for the control of
   ticks on livestock. J S Afr Vet Med Assoc 37 (4):432 – 438
Fink DE (1927) Is glutathione the arsenical receptor in insects? Biochem J 15:296 – 305

Fluck V, Rufenacht K (1969) Effectiveness of newer phosphorous compounds against resistant ticks *Boophilus microplus*. Vet Pestic Soc Chem Ind Monogr 33:183 – 196

García Mata E (1947) Empleo del DDT en la lucha contra la garrapata común del ganado bovino (*B. microplus* (Can)). Minist Agric Gan Nacl Dir Gen Ganadería. Dir San Anim Bs As

Geigy JR, Geigy SA (1962 – 1963) Diazinon contra los ectoparásitos de los animales domésticos. Inf Dep Prop Antiparasit, Basilea, Suiza, pp 1 – 10

Gothe R, Hartig M (1976) On the ixodicidal activity of clenpyrin, chlordimeform and chlormethiuron against OP-resistant *Boophilus microplus* strains. Zur ixodiziden Wirksamkeit von Clenpyrin, Chlordimeform und Chlormethiuron gegen P.E. resistente *Boophilus microplus*-Stämme. Zentralbl Vet Med B23:243 – 254

Graham OH, Drummond RO (1964) Laboratory screening of Insecticides of the prevention of reproduction of *Boophilus* ticks. J Econ Entomol 57 (3):335 – 339

Graham OH, Drummond RO, Diamant G (1964) The reproductive capacity of female *Boophilus annulatus*. Collected from cattle dipped in arsenic of coumaphos. J Econ Entomol 57 (3):409 – 410

Grannet P, Shea WO (1955) Laboratory tests of diazinon-butoxy polypropylene glycol residues. J Econ Entomol 48:487 – 488

Grillo Torrado JM, Gutierrez RO (1977) Susceptibilidad de la metaninfa de la garrapata *Boophilus microplus* can. Lah. frente a los acaricidas organofosforados. Med Vet Rev Med Vet 52 (2):1 – 10

Grillo Torrado JM, Perez Arrieta A (1973) Acción in vitro de los garrapaticidas organofosforados sobre la evolución de los huevos de la garrapata *Boophilus microplus* can. Lah Rev Invest Agrop INTA 6:211 – 221

Grillo Torrado JM, Gutierrez RO, Perez Arrieta A (1971) Comparación de la actividad in vitro e in vivo de los garrapaticidas organofosforados. Rev Invest Agropecu Ser 4 (Pathol Anim) 8:no 3, 59 – 70

Guyton AC (1963) Tratado de fisiología médica. Ed Interam Ed 2:218 – 219

Hammant CA (1977) The introduction of dioxathion for cattle tick control in the tribal trust lands of Rhodesia. Rhod Vet J 8:67 – 70

Harrison IR (1969) The testing of compounds against the cattle tick *Boophilus microplus* with particular reference to the development of 2-cyclohexylphenyl-N-methylcarbamate. Proc 2nd Int Congr Acarol, (Sutton Bonington) 509 – 514

Harrison IR, Palmer BH, Wilhshurst EC (1973a) Chemical control of cattle ticks. Resistance problems. Pestic Sci 4:531 – 542

Harrison IR, Kozlik A, McCarthy JF (1973b) 1,5-Di(2,4-dimethyl-phenyl)-3-methyl-1,3,5-triazapenta-1,4-diene, a new acaricide active against strains of mites resistant to organophosphorus and bridged diphenyl compounds. Pestic Sci 3:679 – 680

Hitchcock LF, Mackerras IM (1947) The use of DDT in dips to control cattle tick in Queensland. J Cun Sci Ind Res Aust 20:43 – 45

Hitchcock LF, Roulston WJ (1955) Arsenic resistance in a strain of cattle tick (*Boophilus microplus* can.) from northern New South Wales. Aust J Agric Res 6:666 – 670

Inoue Y, Ohno S, Mizuno T, Yura Y, Murayama K (1976) Insecticidal activities of synthetic pyrethroids. In: Synthetic pyrethroids. Robert F Gould (ed) Elliot M

Jara de la F (1971) Algunas notas sobre garrapatas del ganado en Mexico (con claves para identificación de géneros y especies) y su combate con Supona. Cienc Vet XVI 2:131 – 153

Jolly DW, Ratcliffe BD (1958) A field method for measuring blood cholinesterase of animals. Vet Rec 3:29 – 31

Jung HF (1959) A new phosphoric ester residual insecticide with a low order of toxicity. Bull WHO 21 (2):215 – 221

Kato T, Veda K, Fujimoto K (1964) New insecticidally active chrysanthemates. Agric Biol Chem 28:914 – 915

Kenaga EE, Whitney WK, Hardly JL, Doty AE (1965) Laboratory tests with dursban insecticide. J Econ Entomol 58 (6):1043 – 1050

Kirkwood S, Phillips PI (1946) Effect of BHC isomers on yeast growth. J Biol Chem 163:251 – 254

Kirshnan KS, Raghavan NG, Mammen ML (1957) Comparative field trials on the residual effectiveness of DDT, malathion and diazinon. Indian J Malariol 12:43 – 48

Knowles CO, Roulston WJ (1972) Antagonism of chlorphenamidine in larvae of the cattle tick *Boophilus microplus* by formamidines and related compounds. J Aust Entomol 11:349 – 350

Knowles CO, Wilson JT, Schnitzerling E (1973) Biossay of chlorphenamidine using larvae of the cattle tick *Boophilus microplus*. Aust Vet J 49:205 – 206

Larkin PJ (1961) Control of the blue tick (*Boophilus decoloratus*) on cattle with pyrethrum sprays. Vet Rec 73:298 – 300

Leahey JP (1979) The metabolism and environmental degradation of the pyrethroid insecticides. Outlook Agric 10:135 – 142

Legg J (1953) A further note on the control of the cattle tick *Boophilus microplus* by use of synthetic insecticides. Aust Vet J 29:200 – 201

Legg J (1956) A test of two organic phosphorus compounds, diazinon and malathion, in the control of cattle tick in Queensland. Aust Vet J 3:55 – 60

Lepesme P (1937) Comptes rendus hebdomaires des séances. CR Acad Sci 204:717 – 728

Lombardero OJ, Luciani CA (1980) Efecto del closantel inyectable 5% sobre la oviposición y la eclosión de huevos de *Boophilus microplus* en bovinos naturalmente infestados. 3rd Congr Argent Cienc Vet, Simp sobre closantel

Lombardero OJ, Moriena RA, Racciopi O (1984) Evaluación de la ivermectina como garrapaticida en bovinos a campo de la Provincia de Corrientes. Rev Med Vet Buenos Aires 1:46 – 50

Loomis EC, Nooderhaven D, Roulston WJ (1972) Control of the southern cattle tick by pour-on animal systemic insecticides. J Econ Entomol 65 (6):1638 – 1641

Ludwig PD, McNeill JC (1966) Results of laboratory and field tests with dursban insecticide for mosquito control. Mosq News 26 (3):344 – 351

March RB (1956) Fat of P-32-labelled malathion sprayed on Jersey heifer calves. J Econ Entomol 49:679 – 682

Martin SIS (1975) A laboratory technique for the evaluation of compounds applied to cattle for the control of the stable fly, *Stomoxys calcitrans*. Proc 8th Br Insectic Fungic Conf, pp 539 – 545

Matter JJ, Schmidt CD, Blumer R (1976) Compounds screened as animal protectant sprays at Kerrville Texas. New Orleans, LA USA, ARS-S 125:1 – 65

McCullough RN, Nolan J (1974) Studies of Acaricides at the cattle tick research station Wollongbar 1963 to 1970. Sci Bull 85:12 – 15

Metcalf RL (1955) Organic insecticides, their chemistry and mode of action. Wiley Interscience, New York, chap. 11, pp 251 – 315

Metcalf RL, Kearns CW (1945) Drug antagonism of DDT: *Periplaneta blatta*. Natl Res Counc, Ins Control Comm, Rep CC 1:109 – 118

Ministerio de Agricultura de la Nacion Argentina (1941) Dir Ganadería Ley 12.566, Bol no 8 April

Ministerio de Ganadería y Agricultura (1941) Dir Ganadería Ley 9.965, Erradicación de la Garrapata. Montevideo Cartilla 1:43 – 44

Muñoz-Cobeñas ME, Moltedo HL, Moiso AM (1978) Sarna psoróptica en ovinos, tratamiento de campo empleando Amitraz. Gac Vet Buenos Aires 40:191 – 196

Munson SC, Yeager IF (1947) Promortem survival times with cyanides. Ann Entomol Soc Am 40:475 – 488

Negherbon WO (1959) Handb Toxicol Nat Acad Sci Nat Res Counc, vol III. Insecticides. Saunders, Philadelphia London, pp 25 – 40

Newton LG (1967) Acaricide resistance and cattle tick control. Aust Vet J 43:1 – 7

Nolan J (1979) New acaricides to control-resistant ticks. Recent Adv Acarol 11 (2):55 – 64

Nolan J, Roulston WI, Wharton RH (1977) Resistance to synthetic pyrethroids in a DDT-resistant strain of *Boophilus microplus*. Pestic Sci 8:1 – 3

Nolan J, Roulston WJ, Schnitzerling HJ (1979) The potential of some synthetic pyrethroids for control of cattle tick (*Boophilus microplus*). Aust Vet J 55:463 – 466

Nolan J, Bird P, Schnitzerling HJ (1981) Evaluation of the potential of systemic slow release chemical treatments for control of the cattle tick (*Boophilus microplus*) using ivermectin. Aust Vet J 57:493 – 497

Norris KR (1956) Commonw Sci Indust Res Org. Research on cattle tick. Aust Vet J 32:177 – 182

Norris KR, Roulston WJ, Snowball GJ (1950) Observations on the control of the cattle tick preparations of DDT and benzene hexachloride (BHC) in dips. Aust J Agric Res 1:165 – 177

Nuñez JL (1977) The history of sheep scab in Argentina. Proc Br Crop Protect Conf Brighton, UK, pp 409 – 413

Nuñez JL, Pugliese ME, Shaw RD (1972) *Boophilus microplus* Can. pruebas de susceptibilidad in vitro con veinte cepas argentinas. Rev Med Vet 52 (1):37 – 43

Nuñez JL, Pugliese ME, Bachmann H (1975) Prueba en gran escala empleando un garrapaticida en base a ethion. Gac Vet Buenos Aires 298:183 – 187

O'Brien RD (1960) Toxic phosphorus esters, chemistry metabolism and biological effects, vol I. Academic Press, London New York, 19 – 27

Pacheco Torres MF (1968) Lucha contra ectoparásitos que afectan la ganadería en Venezuela. Comportamiento de los insecticidas utilizados. Rev Vet Venez 21:202 – 223

Palmer WA (1976a) Acaricidal biossay for cattle tick (*Boophilus microplus*). NSW Dep Agric Tech Bull 12:1 – 12

Palmer WA (1976b) Management of ethion chlordimeform mixtures in dipping baths. NSW Dep Agric Tech Bull 9:1 – 11

Palmer WA, McCarthy JF, Kozlik A (1971) A new chemical group of cattle acaricides. Proc 3rd Int Congr Acarol 2:687 – 691

Pascoe R, Bentley PD, Cheetham R, Hugg RK, Sayle JD (1980) Fluorinated analogues of chrysanthemic acid. Pestic Sci 11:156 – 164

Pasquier R (1947) Symptoms of BHC poisoning: *Schistocerca*. Bull Semestr. Natl Anti-Acaricidien Algeria 4:5 – 22

Pizarro IC, de Cerineo JM, Bocalandro AG, Grimaux CA (1963) Los garrapaticidas fosforados, investigación de su actividad frente al *Boophilus microplus*. Rev Med Vet Buenos Aires 44 (1):13 – 23

Potter C (1935) An account of the constitution and use of an atomized white oil – pyrethrum fluid – to control *Plodia interpunctella* HB and *Ephestesis elvtella* HB in warehouses. Ann Appl Biol 4:765 – 805

Purchase HS (1955) Some thoughts on ticks and their practical control. Bull Epizoot Dis Afr 3:226 – 230

Radeleff RO (1958) The toxicity of insecticides and herbicides to livestock. Adv Vet Sci 4:265 – 271

Reeves GI (1925) Toxicity of arsenical compounds in cattle, sheep and horses. J Econ Entomol 18:82 – 87

Richards AG, Cutkomp LK (1945) Cholinesterase of nerve: *Apis* and *Periplaneta*. J Cell Comp Physiol 26:57–61

Richards AG, Cutkomp LK (1946) Chitinous cuticles and DDT. Biol Bull 90:97–108

Roe RT (1969) The toxicity to cattle of some acaricides in use in plunge dips in New South Wales. Aust Vet J 45:332–337

Roeder KO, Weiant EA (1948) DDT, sensory nerves and campaniform sensilla: *Periplaneta*. J Cell Comp Physiol 32:175–186

Roncalli RA (1980) Evaluación clínica de las avermectinas, una nueva clase de agentes anti-parasitarios de amplio espectro. Mem 3rd Congr Argent Cienc Vet, Buenos Aires, Argentina, pp 182–189

Roulston WJ (1956) The effects of some chlorinated hydrocarbons as systemic acaricides against the cattle tick *Boophilus microplus* Can. Aust J Agrric Res 7:608–624

Roulston WJ, Wharton RH (1967) Acaricide tests on the Biarra strains of organophosphorus-resistant cattle tick *Boophilus microplus* from Southern Queensland. Aust Vet J 43:129–134

Roulston WJ, Wilson JT (1965) Chemical control of the cattle tick *Boophilus microplus* Can. Bull Entomol Res 55 (4):617–635

Roulston WJ, Schuntner CA, Schnitzerling HJ (1965) Metabolism of coumaphos in larvae of the cattle tick *Boophilus microplus*. Aust J Biol Sci 19:619–633

Roulston WJ, Schnitzerling HJ, Schuntner CA (1968a) Acetylcholinesterase insensibility in the Biarra strain of the cattle tick *Boophilus microplus* as a cause of resistance to organophosphorus and carbamate acaricides. Aust Biol Sci 21:759–767

Roulston WJ, Stone FG, Wilson JT, White LI (1968b) Chemical control of an organophosphorus and carbamate resistant strain of *Boophilus microplus* Can. from Queensland. Bull Entomol Res 58:379–392

Roulston WJ, Wharton RH, Schnitzerling HJ, Schustherst RW, Sullivan NO (1971) Mixtures of chlorphenamidine with other acaricides for the control of organophosphorus-resistant strains of cattle tick *Boophilus microplus*. Aust Vet J 47:522–528

Said MS, Atef M, Elrefai AH, Michael S, Elsadr H (1971) Experiments on Asuntol and Bercotox for tick control. J Egypt Vet Med Assoc 31:43–54

Schechter MS, Green S, La Farge FB (1949) Constituents of pyrethrum flowers. XXIII Cinerolone and the synthesis of related cyclopentenolones. J Am Chem Soc 71:3165–3173

Schmidt CD, Matter JJ, Meurer JH, Reeves RE, Shelley BK (1976) Evaluation of a synthetic pyrethroid for control of stable flies and horn flies on cattle. J Econ Entomol 69:484–486

Schuntner CA, Thompson PG (1978) Metabolism of C_{14} Amitraz in larvae of *Boophilus microplus*. Aust J Biol Sci 31:141–148

Schuntner CA, Thompson PG (1979) Toxicology and metabolism of chlormethiuron in *Boophilus microplus* larvae. Pestic Sci 10:519–526

Schuntner CA, Schnitzerling HJ, Roulston WJ (1972) Carbaryl metabolism in larvae of organophosphorus and carbamate susceptible and resistant strains of cattle tick *Boophilus microplus*. Pestic Biochem Physiol 1:424–433

Schuntner CA, Roulston WJ, Wharton RH (1974) Toxicity of piperonyl butoxide to *Boophilus microplus*. Nature (London) 249:386–387

Sergent E, Donatien A, Parrot L (1945) DDT for brown dog tick. Arch Inst Pasteur Alger 23:249–259

Shaw RD (1965) Culture of an organophosphorus resistant strain of *Boophilus microplus* Can. and an assessment of its resistance spectrum. Bull Entomol Res 56:389–405

Shaw RD, Baker JAF (1966) The in vitro activity of Supona against ticks. Vet Rec 25:867–870

Shaw RD, Cook M, Carson RE (1968) Developments in the resistance status of the southern cattle tick to organophosphorus and carbamate insecticides. J Econ Entomol 61:1590 – 1593

Sherman M (1948) Effects of gamma and other isomers of BHC. J Econ Entomol 41:575 – 583

Simms BT (1946) Control of cattle tick and lice. Rep Chief, Bur Anim Indust USA (1945 – 1946):50 – 60

Slade RE (1945) The gamma isomers of hexachlorocyclohexane (Gamexane) an insecticide with outstanding properties. Chem Indust 40:314 – 319

Smallman BN, Fisher RW (1958) Effect of anticholinesterase on acetylcholine levels in insects. Can J Biochem Physiol 36:575 – 586

Smallman BN, Schuntner CA (1972) Authentication of the colinergic system in the cattle tick *Boophilus microplus*. Insect Biochem 2:67 – 77

Stendel W, Andrews P (1973) The development of a new compound active against resistant ticks. Proc 7th Br Insectic Fungic Conf, pp 281 – 289

Stone BF (1968) Brain cholinesterase activity and its inheritance in cattle tick *Boophilus microplus* strains resistant and susceptible to organophosphorus acaricides. Aust J Biol Sci 21:321 – 330

Stone BF, Knowles CO (1973) A laboratory evaluation of chemicals causing the detachment of the cattle tick *Boophilus microplus*. J Aust Entomol Soc 12:165 – 172

Stone BF, Meyers RAJ (1957) Dieldrin-resistant cattle ticks *Boophilus microplus* Can. in Queensland. Aust J Agric Res 8:312 – 317

Stone BF, Webber LG (1960) Cattle ticks, *Boophilus microplus*, resistant to DDT, BHC, dieldrin. Aust J Agric Res 11 (1):105 – 119

Stubbs VK, Wilshire C, Webber LG (1982) Cyhalothrin. A novel acaricide and insecticidal synthetic pyrethroid for the control of the cattle tick (*Boophilus microplus*) and the Buffalo fly (*Haematobia irritans exigua*). Aust Vet J 59:152 – 155

Suarez MF, Serrano MA (1963) Consideraciones a un trabajo técnico sobre garrapaticidas organofosforados. Gac Vet Buenos Aires 156:293 – 298

Thorburn JA (1947) The control of ectoparasitic infestation of farm stock with "Gammexane". With special reference to the arsenic-resistant blue tick. Emp J Exp Agric XV 57:43 – 50

Thorburn JA (1968) Control of cattle ticks with Supona. Rhodesia Agric J 65 (5): 107 – 111

Tobías JM, Kollros JJ, Savit J (1946) Contacts vs. internal action of DDT. J Pharmacol Exp Ther 86:287 – 293

Ulmann E (1972) Lindane monograph of an insecticide. Shilinger, Freiburg im Breisgau, pp 17 – 25

Uribe LF, de Souza LAM, Rae DG (1976) Activity of a new ixodicide Triatix against the cattle tick *Boophilus microplus* under normal field conditions. Argent Fac Vet UFRGS Porto Alegre 4 – 5:122 – 134

Vickery DS, Arthur BW (1960) Animal systemic activity, metabolism and stability of C-Ral (Bayer 21/199). J Econ Entomol 53 (6):1037 – 1047

Voegtglin C, Dyer HA, Leonardo CS (1923) On the mechanism of the action of arsenic upon protoplasm. US Publ Health Rep 38:1882 – 1912

von Orelly IC, Boray IC, Immler R, Knuse F, Wettstein K (1978) The effect of dipofene on resistance ticks. Acta Congr Mundial Vet, Salónica, Grecia

Wade LL (1968) The efficacy and stability of Dursban insecticide in dipping vat for control of the southern cattle tick. J Econ Entomol 61 (4):908 – 909

Weighton DM, Kerry JC, McCarthy JF (1973) Amitraz, a novel acaricide with selective insecticidal properties. Proc 7th Br Insectic Fungic Conf, pp 703 – 711

Welch P, Vincent ST (1975) Ac. 84, 633 (Nimidane): Control of Biarras strain of the cattle tick *Boophilus microplus* on cattle by dipping in Australia. American Cyanamid Co. Int Agric Res Dev Rep 31:374–379

Welsh JH, Gordon HT (1947) DDT on nerve axon of Roach and Cryfish. J Cell Comp Physiol 30:147–171

West TF, Campbell GA (1946) Acaricide resistance and cattle tick control. Aust Vet J 43:394–398

Wharton RH, Roulston WJ (1970) Resistance to ticks to chemicals. Annu Rev Entomol 15:381–404

Wharton RH, Roulston WJ (1977) Acaricide resistance in *Boophilus microplus* in Australia. Workshop on hemoparasites (anaplasmosis and babesiosis Colombia I:73–92)

Wharton RH, Roulston WJ, Utech KBW, Kear JD (1970) Assessment of the efficiency of acaricides and their mode of applications against the cattle tick *Boophilus microplus*. Aust J Agric Res 21:985–1006

Whitnall ABM, Bradford B (1947) An arsenic-resistant tick and its control with gammexane dips. Bull Entomol Res 43:353–372

Whitney WK, Harrison RP, Howe RG (1967) Cockroach control with dursban insecticide. Pestic Control 35:25–30

Wilson RG (1978) Biochemical mechanisms causing tick resistance. J S Afr Vet Assoc 49 (1):49–51

Wood JC (1967) Developments in control of ectoparasites of livestock. Chem Indust:1731–1736

Woodward GT (1957) The treatment of organic insecticide poisoning with atropine sulphate and 2 PAM. Vet Med 52:571–578

VI Tick Control off the Host

The methods which have been discussed up to now are based on the application of chemicals to the infested cattle by means of two systems: dip baths and spraying; that is, eliminating the different parasitic stages when on the host.

In a study of the life cycle of the tick, it can be observed that one part takes place off the host, directly in the environment; this stage is called the free-living or non-parasitic stage.

The examination of these stages makes it possible to find a second way to break the parasitic cycle.

The purpose which encouraged the development of alternative methods for tick control off the host, dispensing with the use of chemicals, can be summarized as follows:

1. Diminishing the possibility of development of strains resistant to ixodicides.
2. Eliminating the risk of environmental contamination and of residues which are left in milking animals and on those to be slaughtered.
3. Reducing the costs arising from dip baths.
4. Provoking a true reduction of non-parasitic elements from the *Boophilus microplus* life cycle.

First, we shall deal with the movement of animals from pastures, or rotational grazing, sometimes also called "pasture spelling".

The first attempts probably date from 1911, when Laws and Manning proposed a control system based on the death of ticks either by eradication of the free-living stages of *Boophilus microplus* or by the absence of available hosts through closing stockyards for at least 2 years.

In 1968 Johnston et al. succeeded with this system in eradicating *Boophilus microplus* from an Australian island, by withdrawing all the cattle and horses for a period of 6 months. Unfortunately, they could not exactly determine which was the minimum period of time necessary to ensure the complete destruction of ticks under field conditions.

In Argentina, an informative booklet from the Ministry of Agriculture (Joan 1940) suggested to herdsmen the temporary closing of pastures for the same ends. But these methods, although effective, are unprofitable because the sources of forage cannot be adequately exploited.

In order to minimize these unprofitable aspects, and taking into account the seasonal variations undergone by the free-living stages, the Australians developed combined systems: rotating periods of pasture together with dip baths, succeeding in reducing the use of ixodicides, to a great extent.

The influence excercised by the rotation of pastures on the decrease in the toll of ticks under field conditions is outstanding. This fact is corroborated in work by Norris (1957, 1956) who, with a precise knowledge of the bioecology of *Boophilus microplus* and using observations made by Snowball (1957), worked with two herds. One herd was subjected to permanent pasture and to spraying each time the parasitic toll per animal increased. The sprayings were repeated eight times during 15 months. Meanwhile, during the same period, Norris subjected the other herd to grazing on one pasture by halves alternately each 3 months, and he needed to spray this herd once only.

Wilkinson (1963) repeated these experiences and achieved better results, not needing to treat the herd on pasture rotation. Furthermore, in order to confirm these discoveries, he exchanged the systems by moving the herd from permanent pasture to a rotational one, and succeeded in decreasing the parasites on this herd to one-third of the previous toll.

Harley and Wilkinson (1964), with slight variations, also tested three herds of Shorthorn cattle during a 2-year period to compare the efficiency of the following routines:

A. Treatment according to degree of infestation.
B. Set treatment every 21 days.
C. Pasture rotation.

The high efficiency of the last two methods arises from the outstanding decrease of the parasitic burden on animals, 79% corresponding to system B and 64% to C. With system C, the number of treatments was reduced by 60%.

Wharton et al. (1969) tried a method similar to that used by Harley and Wilkinson but with two herds, which were a cross between Zebu and British cattle. These were treated by the methods above in order to find out if their natural resistance to ticks would be sufficient to achieve control under conditions favourable to reproduction and survival of the species in humid climates.

Apart from coinciding with the results of the previous experiment, the crossbred herds had only a low parasitic burden and, though their resistance was not sufficiently strong to completely eliminate the necessity of treatment, one of the herds which stayed in a highly infested lot of land required treatment only once in 21 months.

Harley and Wilkinson (1971), applying the results of previous studies carried out by Harley (1966) on the bioecology of ticks, based another trial on the premise that the system of rotating pasture has to be long enough to ensure complete tick-drop off the animal, and short enough to avoid reinfestation caused by hatched larvae belonging to these stages.

To that end, they arranged their trials making use of disinfested pieces of land where the animals stayed for a short period of 6 days in order to avoid their own reinfestation. They were then moved to better grazing until the infestation of ticks reached an unacceptable level, when the rotation was started again.

Although the authors made use of treatment on occasions, this system greatly diminishes the use of ixodicides. Waters (1972) successfully applied a similar system in Queensland for controlling infestation by a resistant tick. Nevertheless, a detailed knowledge of the ecology of the tick is required to apply these systems in different areas.

Ivancovich (1979), for example in the Northwest of Argentina, used Australian methods of permanent and rotating pastures alternating with treatment before entering a new piece of land each time the level of parasitism required it. He succeeded in reducing the number of dips required by 50% in the second system as compared to the first one. He stressed at the same time the biological importance of not losing premunity against anaplasmosis and babesiosis, while decreasing the parasitism level in the country.

Grazing Treatments with Acaricides

Although acaricides were successfully applied in small areas, as described by Ebeling and Pence (1958), who used different compounds such as lindane, sodium arsenite, chlordane and dieldrin, (with best results against ants of the genus *Reticulitermes hesperus*), their use is inapplicable for the control of ticks over large areas. It is also known that organochlorines may also cause toxicity problems in mammals and important alterations in the ecologic balance by the destruction of useful insects.

Repellent Action of Certain Pastures

The latest discoveries on indirect control of ticks, that is, dispensing with the use of ixodicides, are related to the study of types of tropical and subtropical legumes belonging to *Stylosanthes* spp.

The leaves of this nutritious variety of pasture are characteristicallly covered with glandular hairs which secrete a viscous fluid. Sutherst et al. (1982) proved by his experiments that this sticky secretion causes the immediate paralysis of *Boophilus microplus* neolarvae, as they try to climb the grass stalks in order to wait for their future hosts. Additionally, the larvae are intoxicated for 24 h with a vapour, not yet chemically identified, which arises from these secretions.

The author believes that these highly productive legumes seem to offer great potential for the reduction of non-parasitic populations of all tick species, in large tropical and sub-tropical areas where this variety of grass can grow as pasture.

Biological Control by Means of Predators

Birds

Wilkinson (see Wharton et al. 1969) quotes the existence, in Australia, of birds of the *Grallina cyanuleuca* G and *Sturnus vulgaris* L. spp., which act as predators. They have been seen near infested cattle eating ticks that have dropped off.

Nevertheless, from the practical point of view they cannot be considered of great use, as they appear irregularly; this system, according to Wharton and Norris (1980), would be unacceptable in areas where campaigns against ticks take place and they could also represent potential competition to the indigenous species in the ecological balance.

Parasites

Only one parasite belonging to the genus *Hymenopterus* spp. has been found which was in three-host ticks. For these parasites to be used, they would first have to be adapted to *Boophilus microplus*, and their real degree of pathogenicity would have to be authentically proven (Wharton and Norris 1980).

Bacteria and Viruses

Provided a pathogenic strain were available, periodic applications to cattle in the same way as acaricides would be required, and this would be likely to lead to the selection of resistance in a similar way.

Ants and Spiders

Very few species of truly predatory arthropods have been found up to the present time, although the spider *Lycosa go deffray* (Koch), has been seen carrying ticks on its back to its nest. Ants have also been found within tick areas, and are thought to feed on the eggs laid by engorged females, generally in spring time. It is likely that, during the other seasons, their feeding habits vary according to the availability of other kinds of food.

The most common species of predatory ants are as follows: *Pheidole megacephala; Pheidole weisei; Arphaenogaster longiceps; Chalcoponera metallica; Iri-*

domyrmex mjobergi; Iridomyrmex anceps; Thytidoponera cristata; Meranoplus hirsutus and *Acrocoflia australis.*

Another problem which exists in their use is that sufficient knowledge of appropriate types of soils for the establishment of their nests is not available, and it is not possible to foresee the intensity of their attack, which probably varies in accordance with some unidentified physiological requirements.

Wasps

Wharton and Norris (1980) have indicated that larvae and nymphs of three-host ticks have been attacked by small wasps, but nevertheless do not feed on them exclusively. No such attack on *Boophilus microplus* has yet been recorded.

Sexual Sterilization of Adult Males

Systems for the production of sterile males, as used for the control of insects — as Waterhouse (1973) points out — have no practical application for the control of *Boophilus microplus.*

The number needed to achieve a ratio of 10 to 1 of sterile males to fertile ones would be very high. This would be uneconomic because ticks are strictly parasites and require the presence of a host to fulfill their life cycle; numerous laboratories of tick donor animals would be necessary. Such attempts as have been carried out in vitro to develop the life cycle of *Boophilus microplus* have had little success (Kemp et al. 1975).

Furthermore, according to Galum et al. (1974), irradiated adult males lose their mobility earlier than fertile ones due to the inhibition of their spermatogenesis.

In Argentina, Guisasola (1969), encouraged by investigations carried out by nuclear methods already developed in several countries on different pests, observed the effects of gamma radiation of Co on oviposition of *Boophilus microplus.* She found that radiations of 500 rad affected the viability of eggs, and by 30%, the eclosion of larvae.

Metabolic Alterations

Stone (1977 personal communication) indicated that great efforts have been made to investigate which substance secreted by ticks participated in the adherence of ticks to the host. Once discovered, the purpose would be to prevent the secretions.

Hormones

The possibility of immunizing cattle against hormones produced by ticks has also been suggested. These hormones are active in moultings from one stage to the next in the life cycle.

Galum et al. (1974) found that cattle can produce specific antibodies against hormones produced by ticks. Those antibodies produced against the action of hormones would be able to inhibit physiological actions either in vivo or in vitro. Galum thinks that the development and maturation of insects, and probably also of ticks is governed by three main hormones: the cerebral, the juvenile and the ecdysone, which induce moulting in arthropods during their life cycle.

Changes in titres between the juvenile hormone and the ecdysone determine the nature of the moulting. For moulting to take place, the titre of juvenile hormone must decrease and that of the ecdysone increase.

Juvenile hormones were chemically characterized and have been used for the production of antibodies in rabbits; if this effect were achieved in cattle, it is presumed that the antibodies would penetrate ticks by way of the blood on which they feed; the antibodies would then pass through the gut wall without being altered in the hemolymph, where interaction with the above-mentioned hormones would finally take place.

Sexual Hormones

Gladney et al. (1974) was able to control *Amblyomma americanum* successfully by using pheromones produced by adult males which attract females of the same species located far from the place of application (1.10 m from the treated area). Basically, a mixture of insecticides and pheromones was applied to the ears of treated animals, thus attracting females which were intoxicated when in contact with the mixture.

This method offers the advantage of greatly reducing the quantity of insecticide used, as well as the residues and the toxicity of the insecticides to the animals.

Use of Resistant Cattle

Based on the natural resistance that certain cattle breeds have to limit the number of larvae which can moult to adults, some authors — mainly Australians such as Kearnan (1977), Waterhouse (1962), Wharton (1974) — have proposed the introduction of this system together with pasture rotation as interesting alternatives in the permanent fight against ticks. This will be fully discussed in the following chapter.

References

Ebeling W, Pence RJ (1958) Laboratory evaluation of insecticide-treated soils against the *western subterranean termite*. J Econ Entomol 51(2):182 – 207

Galum R (1976) Control of livestock pests by means of insects growth regulators. Proc Semin Ectoparasites, Cali, Colombia, pp 142 – 163

Galum R, Warburg M, Sternbert S (1974) In the sterile insect technique and its field application. IAEA, Vienne, pp 24 – 26

Gladney WJ, Grabbe RR, Ernst SE, Oehler DO (1974) The Gulf Coast tick: evidence of a pheromone produced by males. J Med Entomol 11(3):303 – 306

Guisasola AA (1969) Efecto de la radiación con rayos gamma sobre desoves de *Boophilus microplus* (Lahille, 1907). Rev Med Vet Buenos Aires 50(6):475 – 586

Harley KLS (1966) Studies on the survival of the non-parasitic stages of the cattle tick *Boophilus microplus* in three climatically dissimilar districts of north Queensland. Aust J Agric Res 17:387 – 410

Harley KLS, Wilkinson PR (1964) A comparison of cattle tick control by "conventional" acaricidal treatment, planned dipping, and pasture spelling. Aust J Agric Res 15:841 – 853

Harley KLS, Wilkinson PR (1971) A modification of pasture spelling to reduce acaricide treatments for cattle tick control. Aust Vet J 47:108 – 111

Ivancovich JC (1979) Control de la garrapata mediante manejo. INTA Est Exp Agropecu El Colorado (FSA) 4:1 – 10

Joan T (1940) La garrapata común del ganado vacuno. Bol Fomento Ganadero Minist Agr Gan Nac 16:1 – 37

Johnston LAY, Wharton RH, Calaby JH (1968) Eradication of cattle tick (*Boophilus microplus*) from magnetic Island, Queensland in the presence of native fauna. Aust Vet J 44:403 – 405

Kearnan JF (1977) Tick control without dipping. Queensl Agric J 3:101 – 103

Kemp DH, Koudstaal DK, Roberts JA, Kerr JD (1975) Feeding of *Boophilus microplus* larvae on a partially defined medium through thin slices of cattle skin. Parasitology 70:243 – 254

Laws HE, Manning B (1911) Eradication of ticks by the starvation method. Gov Agric J Cape of Good Hope 7:1 – 19

McCullough RN, Barrow J (1970) Effects on production of dipping cattle of control ticks. Agric Gaz N S W 5:295 – 299

Norris KR (1956) AVA Conf Pap, Brisbane 1956. CSIRO, Camberra. Aust Vet J 8:177 – 182

Norris KR (1957) Strategic dipping for control of the cattle tick, *Boophilus microplus* in South Queensland. CSIRO, Camberra, pp 768 – 787

Roulston WJ (1956) The effects of some chlorinated hydrocarbons as systemic acaricides against the cattle tick, *Boophilus microplus* Can. Aust. J Agric Res 7:608 – 624

Snowball GJ (1975) Ecological observations on the cattle tick, *Boophilus microplus Can.* CSIRO, Sidney

Sutherst RW, Jones RJ, Schnitzerling HJ (1982) Tropical legumes of the genus *Stylosanthes* immobilize and kill cattle ticks. Nature (London) 295:320 – 321

Waterhouse DF (1962) Insect control by radiation sterilization in Australia International. J Appl Radiat Isot 13:435 – 439

Waterhouse DF (1973) Pest management in Australia. Nature (London) 246:269 – 280

Waters KS (1972) "Pasture-spelling" for the tick. Queensl Agric J 98(4):170 – 175

Wharton RH (1974) Ticks, with special emphasis on *Boophilus microplus*. Entomol CSIRO Indooroopilly, Queensland, Aust. Reprinted from: al R, Wharton RH (eds) Control of arthropods. Plenum, New York

Wharton RH (1975) Resistant cattle for biological control of cattle tick. Vet Sci Intersect Symp 46th Anzaas Congr Camberra 4(2):13 – 17

Wharton RH, Norris KR (1980) Control of parasitic arthropods. Vet Parasit 6:135 – 164

Wharton RH, Harley KLS, Wilkinson PR, Utech KB, Kelley BM (1969) A comparison of cattle tick control by pasture spelling planned dipping, and tick-resistant cattle. Aust J Agric Res 20:783 – 797

Wilkinson PR (1963) Pasture spelling as a control measure for cattle ticks in southern Queensland. Aust J Agric Res 15:822 – 840

Wilkinson PR (1970) Factors affecting the distribution and abundance of the cattle tick in Australia. Observations and hypotheses. Acarologia 12:492 – 508

VII Host Resistance to Ticks

There is no doubt that acaricide treatments are an important item of expense in cattle breeding, not only the cost of ixodicides, but also labour, time, and especially the losses caused by the movement of cattle: weakening (sometimes considerable in extensive farms), traumatisms, separation of calves from their mothers and other accidents inherent to the treatment; toxicity problems, which, although now less frequent on account of new active principles, should not be excluded as they still appear from time to time in areas where organophosphorous compounds are still applied.

In Argentina, in areas where campaigns to eradicate *Boophilus microplus* are carried out, between 14 and 17 dippings take place each year with organophosphorus compounds, and from 8 to 12 with the more modern drugs, amitraz and synthetic pyrethroids; while in the infested zone where only a moderate control of the parasite is attempted, the number of treatments ranges from 4 to 7 in the same period. Thus, it is easy to calculate the work and cost that this process implies in cattle raising.

When ticks develop resistance to the different chemicals used for their control, this problem becomes more serious, making the work on infested herds extremely troublesome and difficult.

For this reason, other means of fighting *Boophilus microplus* have been and are being tried, for example those described when dealing with the subject of control off the host, applying various chemical and biological systems which sometimes produced encouraging results and sometimes only helped to end a research work. Nevertheless, there remains another factor for controlling *Boophilus microplus*, which we are going to deal with next: *host resistance to the parasite.*

Going back to the first paragraph where we stressed the great expense that acaricide treatments bring in a cattle farm where a rational control of ticks is carried out, it is easy to realize how interesting the control of parasites by these means would be, tending to eliminate the routine treatments or at least reducing their number, thus increasing the possibility of making management more effective and of increasing profitability.

Host resistance to *Boophilus microplus* becomes apparent in different ways, the most outstanding being the following:

1. The tendency of resistant animals to lower infestation than those which are susceptible, that is, those which, under equal conditions, bear considerably heavier parasite loads. Thus, assuming equal potential infestation, the difficulties which parasites encounter in fulfilling their life cycle on cattle belonging to the first group are notorious; a fact which results in the remarkably low number of specimens which, in fact, fulfill their parasitic cycle.
2. The smaller size and the lower weight, as well as the decreased rate of oviposition in those engorged females which succeed in fulfilling their parasitic stage within the life cycle.

These facts gave rise to a great variety of trials and prolonged studies carried out mainly in Australia and also in South American countries — especially in Argentina and Brazil — where observations are being verified and where the discovery of the factors and/or mechanisms which cause this resistance is being attempted.

It is no easy task to collect, arrange and then exhibit the copious and varied material on the subject under discussion, since the theories and proved facts are so numerous that the number of hypothesis as well as the opinions, which are not always relevant to the facts, render the task of compilation highly laborious.

For this reason, the subject has been dealt with as follows:

1. Resistance in zebu (*Bos indicus*) and its crosses with cattle of European origin.
2. Resistance in European cattle (*Bos taurus*) and later, jointly for both species, those factors which seem to have a greater influence upon the manifestation of host resistance such as:

A) Influence of the anatomical and physiological characteristics of the skin.
B) Hypersensitivity.
C) Immunity.
D) Grooming (licking and cleaning).

1 Resistance in Zebu (*Bos indicus*) and its Crosses

The first paper published on this subject in South America was by a Brazilian author, Villares (1941). This paper is based on the comparison between Indian-bred cattle (Nellore, in this case) and native and European cattle, from various tests carried out in different places in Brazil. These tests proved that 85% of the total of engorged females collected came from European cattle, 6.7% from Creole cattle and the remaining 4.75% from Indian cattle.

The size of engorged females also showed considerable differences, thus, for example 17.9% of those collected from European breeds were more than 7 mm long, whereas the percentage of those females from native or zebu cattle was 3.4 mm (average of both).

Villares (1941) came to no definite conclusion with regard to the influence of sebaceous secretion, the characteristics of the hair or thickness of skin as factors which cause zebu resistance to tick.

In 1943 an Australian author, Kelley, published a long and interesting paper on zebu and its cross-breedings in Northern Australia. Having established the origin and distribution of the different bovine species, the author stresses the outstanding adaptation of *Bos indicus* to tropical and subtropical areas, and also mentions some of its characteristics which are directly related to our subject, such as repellence of ectoparasites: flies (buffalo fly) and ticks (*Boophilus microplus*).

Table 7.1. Length, width, weight and oviposition of engorged females collected from zebu, European cattle and crossbreeds

Breed	Tick measurements	Length (mm)	Width (mm)	Weight (g)	Oviposition	
					Maximum	Minimum
Zebu	81	6.5	4	0.055	18	1818
Half-breed	140	8.3	5.4	0.136	189	2532
European	270	10.6	7	0.249	782	3610

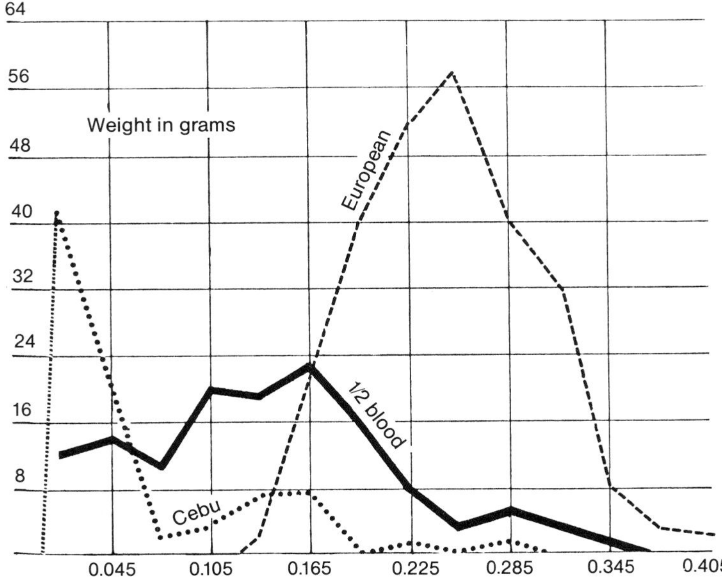

Fig. 7.1. Weight (g) of engorged females obtained from Zebu, European breeds and half-blood cattle. Lombardero and Schiffo (1968)

In resistance to ixodide Kelley attributes an important role to sebaceous secretion and to the skin consistency which, although thin, is dense and hard to penetrate, as well as to the development of the cutaneous muscle, which allows *Bos indicus* to move its hide easily, and also to the great mobility of the tail and ears.

Kelley goes further when he considers the zebu to be completely repellent to ticks, asserting that these characteristics are seen in the cross-breedings in direct proportion to the percentage of zebu blood.

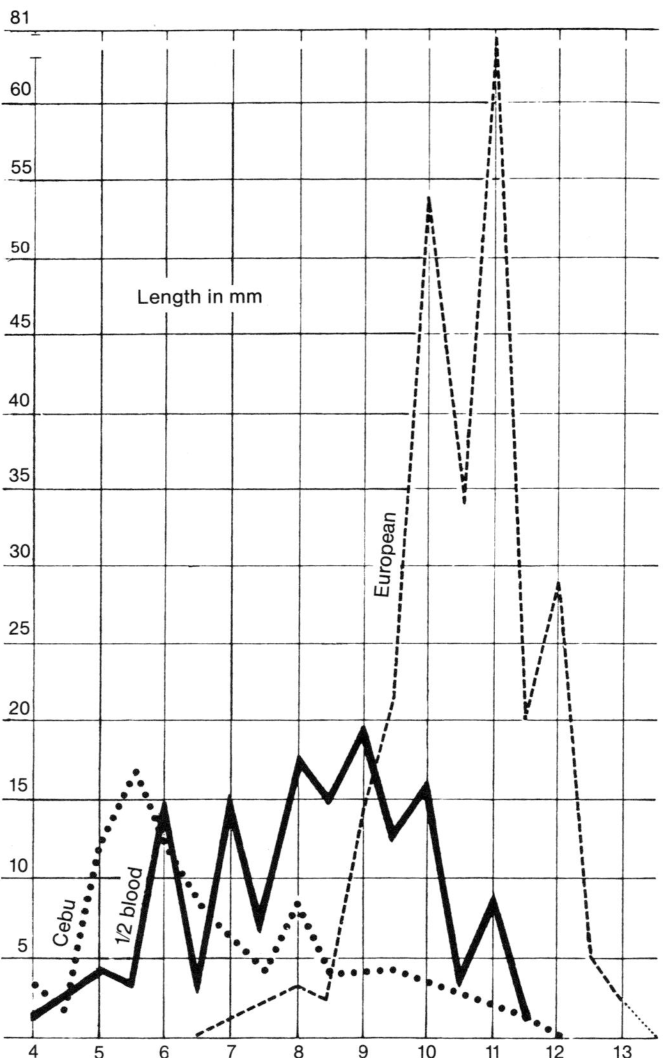

Fig. 7.2. Length (mm) of engorged females collected from Zebu, European breeds and half-blood cattle. Lombardero and Schiffo (1968)

Finally, this investigator asserts that in animals with 25% zebu blood resistance seems to be related to the length and characteristics of the hair: when cattle show their zebu ascendancy (that is, short, straight and shiny hair), repellency is like that of a half-breed zebu; on the contrary, when the hair is longer, of European type, the level of infestation is higher, but does not reach the level of parasitism of pure-bred European cattle.

In Argentina, Lombardero and Schiffo (1968) compared some characteristics of ticks which parasitize *Bos indicus* and *Bos taurus* as well as half-breed animals (first cross), as regards length, width, weight and oviposition of engorged females.

Those differences, shown in Table 7.1, become evident on analyzing Figs. 7.1, 7.2 and 7.3.

The same authors, in a later paper (1970), observe that parasitism occurs in inverse relation to the proportion of zebu blood and they also emphasize that

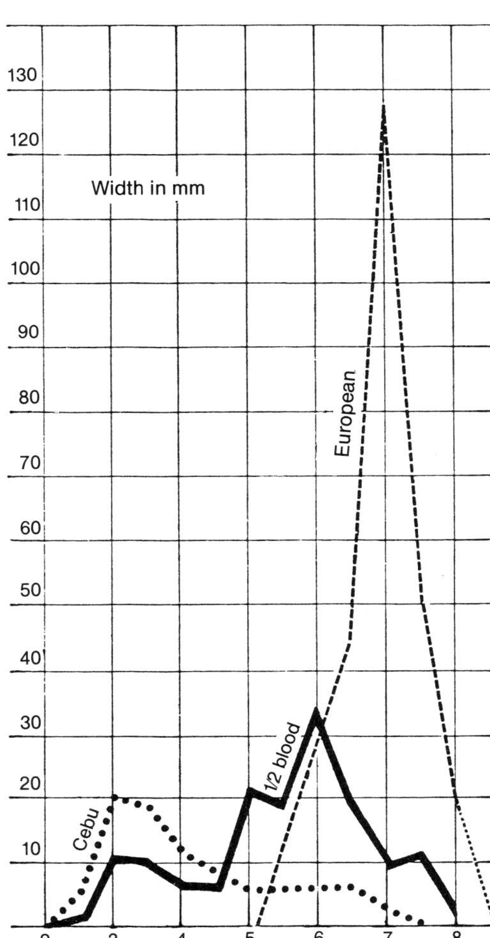

Fig. 7.3. Width (mm) of engorged females collected from Zebu, European breeds and half-blood cattle. Lombardero and Schiffo (1968)

tickicide baths in fact make zebus much more susceptible; they generally recover their natural resistance, only several months later.

Riek (1956), after carrying out several trials by crossing *Bos taurus*, *Bos indicus* and their cross-breeds, believes that there is one, innate resistance, occurring much more frequently in cattle of Indian breeds than of European breeds; and another, acquired resistance, which, in some cases, can be achieved by infesting the animals repeatedly with *Boophilus microplus*. This author finds no correlation between the degree of resistance and the thickness of the skin and the proportion of sweat and sebaceous glands.

Wagland (1979), another Australian author, made an important contribution to this subject. In one of his experiments he infested a group of Brahman steers with a daily load of 1000 larvae until, several weeks later, he obtained a relatively stable number of engorged females. After slaughtering these steers, he recovered the parasitic population by treating the hides, under heat, with caustic soda; he classified and counted the parasitic population, proving that, fundamentally, host resistance to parasites is shown when they are attached to the skin.

In a later publication Wharton (1975) refers to the marked resistance of zebu and its cross-breeds to ticks, estimating it at about 99% on Brahman specimens

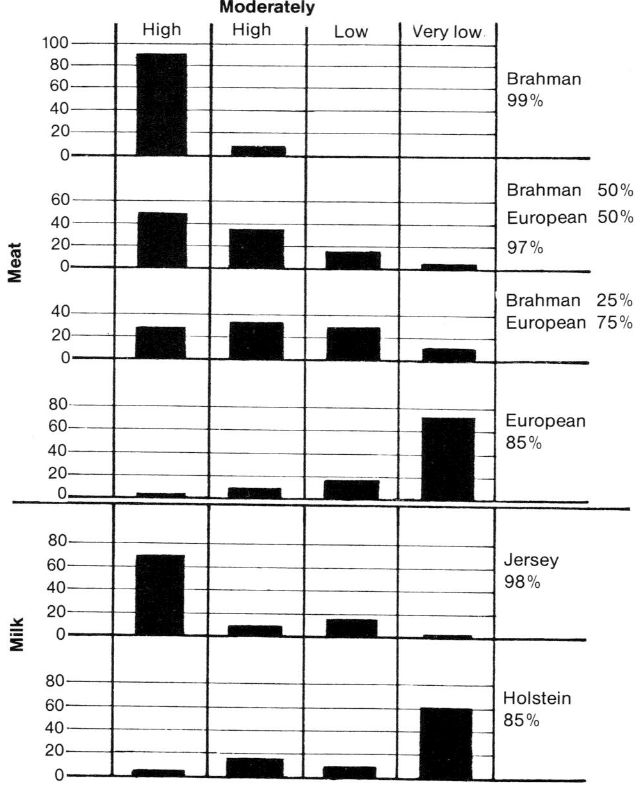

Fig. 7.4. Resistance to *Boophilus microplus* in different cattle breeds. Wharton and Norris (1980)

(this percentage indicates that only 1% of the larvae survived and fulfilled their parasitic cycle). This investigator compared this index to that obtained with half-breed crosses and three-quarter zebu in addition to European breeds which show a much higher susceptibility, except in the case of Jersey cattle, whose particular resistance to ticks will be further dealt with (see Fig. 7.4).

In our professional experience, we have had the opportunity of observing in the whole Northeast area of Argentina as well as in Paraguay and Brazil, the lower infestation suffered by Indian breeds (especially Brahman and Nellore); the smaller size of engorged females collected from these *Bos indicus* breeds in comparison with those collected from Hereford, Shorthorn, Aberdeen Angus and Holstein cattle; the higher susceptibility of Indian cattle when treated with modern acaricides mixed with emulsions obtained by means of strong tensio-active agents (which are also detergents); and the gradual recovery of these cattle after several months, a fact which we will deal with further when considering the possible influence of sebaceous secretion upon this characteristic.

2 Resistance in European Cattle (*Bos taurus*)

In 1908, an Australian farmer, Munro Hill, noticed in Jersey dairy cattle the presence of animals suffering from lesser tick infestations than others belonging to the same breed and under similar conditions; he asserted that this condition may be passively transmitted by inoculating blood from resistant animals. Nevertheless, this last fact could not be proved when studied scientifically with the means available at the time. This was the first report on the existence of a group of European cattle which offered higher resistance to *Boophilus microplus* infestation. It must be taken into account that they were Jersey cattle, a breed which is characterized by a higher proportion of resistant individuals than any other breed of European origin.

Many years later, Riek (1956) contributed plentiful data on the subject and came to the conclusion that only a small proportion of Shorthorn cattle show a significant resistance to common tick infestation, its mechanism being fundamentally of an allergic type.

Kemp et al (1976) studied in detail the larval behaviour and clinical characteristics of infestation on European animals of low and high resistance, trying to elucidate the mechanisms which result in this resistance. It is interesting to see (Fig. 7.5) the description of the effect of host resistance on *Boophilus microplus* larvae.

Modern trials such as those by Wharton et al. 1970, Wharton and Norris 1980 and by Sutherst et al. (1979) agree that although there are resistant individuals in herds of European breeds, the proportion is low, thus rendering the task of selection prolonged and difficult.

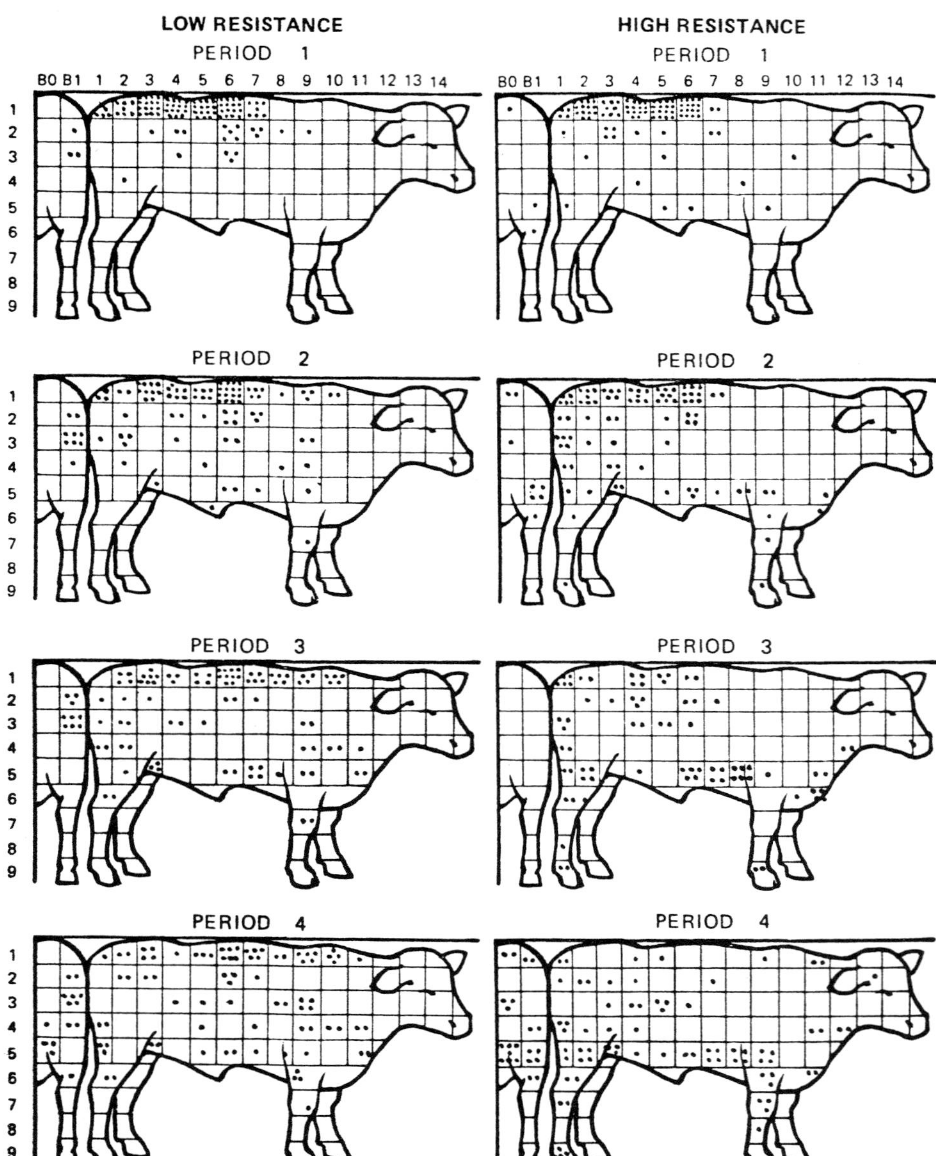

Fig. 7.5. *Boophilus microplus*, the effect of host resistance on larvae attachments and growth. Kemp et al. (1976)

Our experience of resistance in European cattle raised in Argentina, is relative, for we detected resistance only at insignificant percentages; whereas in hundreds of tests carried out on cattle herds of different breeds and ages, we have always noticed a high proportion of susceptible individuals, on very few occasions have we observed low individual infestation on regularly or highly parasitized cattle herds. The proportion of partially resistant animals increased slightly when examining native cattle, especially North of Cordoba, Formosa, East of Salta, South of Santiago del Estero and North of Santa Fe (all of these being Argentine provinces).

The only publication on this subject in Argentina was by Ivancovich (1970), who carried out an experiment in Formosa using 12 bovines which he divided into 3 lots of 4 animals each, in the following way:

Lot A: cross-breed animals (native cattle, Shorthorn, Holstein and Aberdeen Angus) considered resistant.

Lot B: four Holstein heifers considered susceptible.

Lot C: four heifers result of the cross between cross-breed cows (similar to lot A) and Holstein.

The animals were infested with 10,000 larvae each; the following results were obtained after having collected the engorged females from the left flank of the animals.

Lot	*Average per animal (left flank)*
A	22.7
B	231.5
C	10.7

Undoubtedly, the low number of animals used may render the experiment unrepresentative; and the fact that lot A consisted of adult cows, and the others of heifers, may have influenced the results. Nevertheless, the differences obtained were considerable enough to take into account. We also believe that native cattle may have some influence, since — as mentioned above — it is usual to find in this type of cattle or in its cross-breeds a higher percentage of resistant individuals than when working on European breeds.

A Influence of Anatomical and Physiological Characteristics of Skin

The fact that the anatomical characteristics of the skin in Indian cattle favours repellence of ectoparasites, *Boophilus microplus* in this case, has always been stressed.

Kelley (1943), in Australia, and Villares (1941), in Brazil, give fundamental importance to this fact, claiming that mobility of both the skin and of the ears and tail plays an important role in the resistance of zebu cattle. These authors also stress the importance of long hair as compared to short, shiny hair (zebu type), which is much more resistant. In contrast, a higher susceptibility is directly related to a layer of longer hair, such as that of European cattle, even when crossed with zebu. Finally, no importance is attributed to the thickness of the skin.

This opinion is also shared by Riek (1956, 1962), who, although recognizing that Indian and Jersey breeds (the latter having the thinnest skin of all European cattle) bear thinner integuments, ascribes no main function to this characteristic. He stresses that, when studying the skin of both cattle breeds, from the histological point of view, no striking differences are observable as regards the number and type of sebaceous and sudoriferous glands.

In Argentina, Rovere (1977 personal communication) carried out a histological study of the normal skin of Brahman in comparison with that of Hereford and Holstein, also finding no striking differences in its structure, including the sudoriferous and sebaceous glands.

Nevertheless, in 1955 Wilkinson pointed out that there existed a certain relation between the degree of susceptibility and the thickness of skin, although in a later trial published in 1962, he proved that there is some kind of correlation between the resistance to ticks and either the thickness of the skin or the size of the sudoriferous glands; yet he asserted that the depth of the hair follicle plays an important role.

As can be seen, opinions do not always agree, but summing up, we can say that skin anatomy, and length and characteristics of the hair may be of importance in some cases, but do not constitute a defining factor. Nevertheless, the sweat and sebaceous glands, by means of their secretions, may play an interesting part in acarus repellency.

Findlay and McEwan Jenkinson (1960) studied the morphology of sweat glands in cattle; and Yang (1951) carried out histochemical studies comparing secretions of sweat glands between zebu and an European breed, Ayrshire in this case. His main conclusions were the following:

a) The chemical composition of sudoriferous gland secretions of cattle (both European and zebu) differs remarkably from that of human glands, since they contain neither glycogen nor lipids or associated compounds, nor iron.

b) Those reactions carried out so as to confirm the presence of either lipids and cholesterol or the esters thereof gave negative results both in the zebu sweat glands and in those of Ayrshire, yet these substances were found in abundance in sebaceous glands.

All these details suggest that, as no marked differences exist between the sweat gland secretions of animals both susceptible and resistant to *Boophilus microplus*, the product of these glands is of no significance in this regard.

The sebaceous secretion is quite a different case: we have said previously that zebus which had not been dipped showed little susceptibility to common cattle tick infestation and that they may become both partially and temporarily susceptible when dipped with emulsifiable acaricides containing tensio-active agents, which in turn also act as detergents dissolving the sebaceous secretion.

This characteristic is easily observable in zebu-raising areas, where first-cross and three-quarter cattle herds are also numerous. This characteristic has been described by several authors, among them Lombardero and Schiffo (1968) and Lombardero et al. (1970). Apparently, the sebaceous secretion is in some way related to this fact.

Nicolaides et al. (1968) carried out an interesting study by means of thin-layer chromatography on lipids existing on the surface of the human skin and on that of 18 different species of animals, among them *Bos taurus*. In this study, the highest concentration of lipids was observed in monoester and diester zones, and less concentrated in triglycerides.

Miranda Smith et al. (1975, Smith and McEwan Jenkinson 1975, Smith and Ahmed 1976) have worked in this subject over a long period, and their conclusions may be summarized as follows:

a) The sebaceous secretion may be lower in winter than in other seasons.
b) Neither age nor sex significantly affects sebaceous secretion.
c) Temperature, after short-term exposure both to cold and heat, seems to have no influence upon the volume and chemical composition of the secretion.
d) On the other hand, after prolonged exposures to hot temperature, both the volume of secretion and the linoleic acid increase; whereas at low relative humidity, the proportions of palmitic and myristic acids increase.
e) The composition of sebum output both from the skin surface and from the sebaceous glands is as follows:

Lipid fraction	Skin surface lipid (%)	Sebaceous gland lipid
Phospholipid	4.4	12
Unesterified fatty acid	13.7	31.6
Triglycerides	19.5	12.1
Band 1	22.9	18.1
Band 2 and 3	11.4	6
Cholesteryl esters	28.1	20.3

We have not heard of any other comparative studies between secretions of European and Indian cattle breeds, and it is possible that on carrying them out, some striking differences or differences may be detected, which would give us an idea of the repellent action that the zebu and its cattle have to certain ectoparasites. In our case, we consider that terpenoids may play an important role, a hypothesis which is strengthened when the repellent action of some sesquiterpenoids to *Boophilus microplus* is tested in vitro (Matthewson 1979 personal communication).

B Hypersensitivity

On analyzing the causes of resistance to the common cattle tick in certain bovine breeds or individual cattle, the hypersensitivity phenomena − already described by Riek (1956, 1962) − must not be ignored. This Australian author points out that the histological picture in those places where larvae, nymphs or adults, were attached on resistant cattle is characterized by an allergic type reaction, the result of a hypersensitivity response to the salivary secretion of cattle tick.

In the case of *Bos indicus* there would also be − according to Riek − an immunity response which would cause the death of the parasite; whereas, with *Bos taurus*, a greater number of mastocytes on the dermis of the few resistant animals are observed, considering that the sensitization of these cells leads to an antigen-antibody-type reaction which results in the release of histamine which, in turn would cause the edema and irritation that follows the attachment of parasitic elements. In the case of highly resistant cattle, a histamine peak appears in blood 48 h after the attack, and this peak gradually decreases until it is normal, 7 or 8 days later, whereas in the case of susceptible animals no variations in the amount of histamine are observed during the course of the parasitic cycle.

Finally, Riek proves that in resistant animals, the inoculation of 0.01 ml from a dilution of $1:10^6$ histamine phosphatidic acid results in a visible reaction; whereas in susceptible animals, a dilution of $1:10^4$ must be inoculated in order to achieve a similar reaction. He further holds that several serological tests (hemagglutination, complement fixation, etc.) did not show the existence of free antibodies.

On the other hand, Willadsen and Williams (1976) isolated from larvae a protein of a molecular weight of 60,000, which causes an immediate hypersensitivity response in those cattle previously exposed to parasitism.

Later, in 1978, these authors, working with a larvae extract, isolated two allergens which caused this immediate hypersensitivity, and they investigated the relationship of these allergens with the hypersensitivity and the amount of antibodies in the blood, reaching the interesting conclusion that there is an evident correlation between hypersensitivity responses and the level of agglutination to the resistance of resistent cattle or host (Willadsen et al. 1978).

Finally, in 1979, these same investigators (Willadsen et al. 1979) proved that the total amount of histamine on the skin is related in direct proportion to both the level of resistance as well as to the immediate hypersensitivity reactions: treatment with a strong antihistaminic, mepyramine maleate, suppresses the cutaneous hypersensitivity reactions. All this leads to the conclusion that histamine is the principal intervening agent, and that the volume of this substance on the skin may play an important role in resistance of *Boophilus microplus*.

In Argentina, Lombardero et al. (Lombardero and Schiffo 1968, Lombardero et al. 1970) observed on animals of Indian breeds an immediate reaction to the cattle-tick bite (in this case we refer to larvae); and we have also been able to prove, always on Indian breeds, a great activity after infestation (whether natural or artificial) and, in a few cases, a reaction of papular type occurs sometimes accompanied by a small local edema in the adhering zone. These lesions later develop into a cutaneous erythema and disappear.

However, we believe that the hypersensitivity theory needs more proof in order to be accepted without discussion.

C Immunity

It was again Riek (1956) who first made reference to the possibility of developing an immunity process. In a debate which took place in 1956 in Australia, this author gave his opinion that two types of antibody existed: the precipitin and the "hypersensitized antibodies", which can be present simultaneously. Riek (1962), in another trial, said he had obtained positive results by inoculating resistant animals with an antigen prepared with eggs of *Boophilus microplus* in dilution of $1 : 10^3$ or $1 : 10^4$; but he found no correlation between the degree of resistance and the reaction to the intradermic administration of this antigen.

Several years later, Wagland (1979) agrees with Roberts (1968), as well as with Hewetson (1976), that the exact mechanism of host resistance to *Boophilus microplus* is still unknown; but all factors indicate that the process is of an immunity type or that at least, its principal compound is of this nature.

This hypothesis, although not yet proved, is strengthened by investigations by Roberts and Kerr (1976), who maintain that, when the plasma of animals highly resistant to *Boophilus microplus* is transferred to calves, which have not been parasitized, it confers a variable degree of resistance; whereas the plasma from low resistant cattle is not so active when compared with that from uninfested donors. The infestation was carried out with 5000 larvae and the parameter used for calculating the results was based on the number of detached engorged females.

Previously Roberts himself had come to the conclusion that the different degrees of cattle resistance to the parasite were the result of an immunity response. He had proved that approximately 50% of larvae completed their parasitic cycle on the majority of the animals not previously parasitized but that 8 days after infestation, resistance began to show and then each animal always tended to have a constant number of parasites. Hence, the degree of resistance was variable and dependent upon certain individual characteristics of the host; however, the percentage of larvae that completed their cycle ranged between 0 and 30, that is a percentage remarkably lower than that observed in virgin animals. Roberts further noticed that the animals previously parasitized immediately manifested their resistance level. Regarding the duration of the parasitic cycle (21.3 ± 1.6 days), this investigator affirmed that it is not affected by the immunity response.

Finally, Schleger et al. (1976) also suggested the possibility of an immune mechanism, which would result both in the appearance and later lysis of mastocytes, as well as in a marked eosinophyl on the adhering site of parasites in resistant animals. This phenomenon cannot be observed in those animals which have not been previously infested.

Studies on *Dermacentor andersoni*

All the foregoing information arises from trials on *Boophilus microplus* carried out by Australian investigators who worked with cattle. It may therefore appear strange to deal with experimental work on *Dermacentor andersoni* carried out on guinea pigs in Canada. But we include them first, because *Dermacentor andersoni* is an ixodide and thence, from the taxonomic point of view, bears a close relation to *Boophilus microplus*; and second, because these Canadian trials give us extremely important data for interpreting, from the standpoint of immunity, the phenomenon of host resistance to parasites.

Wikel and Allen (1976a), investigators from Saskatchewan University, who have studied the subject in depth, proved that guinea pigs develop resistance to *Dermacentor andersoni* larvae during the first infestation and between 70 and 90% of the larvae engorge, that is they reach a stage similar to that of the engorged larva of *Boophilus microplus*. Yet, after being infested again after 12 days, only between 5 and 15% of the total number of infesting larvae reach this stage. They also observed that the mean weight of larvae significantly decreases (from 22.3 to 2.3 mg) (Allen 1973).

On trying to achieve passive immunization by the transfer of serum these authors obtained no better results, but, on transferring ganglionic lymphocytes, the transfer of passive resistance was achieved, though not to the same degree as that obtained by natural infestation.

Wikel and Allan (1976b), in a complementary publication, proved that the administration of an immunosuppressive, cyclophosphamide (300 mg kg^{-1}) 48 h before infestation partially prevented resistance, which is why these authors think that there exists a humoral component, besides the one already proved and which is apparently, the more important cellular component.

Late in 1978, these same investigators (Wikel and Allen 1978) confirmed the presence of an antigen in the salivary glands of the *Dermacentor andersoni* that causes a quick reaction at skin level in resistant guinea pigs, and this antigen, when cultured in vitro with lymphocytes from resistant hosts, causes blastogenesis.

Already in 1979, Allen et al., using immunofluorescence technique when studying the skin of infested resistant guinea pigs, could observe that some dendritic cells (which they consider Langerhans' cells) combine with the antigen from the salivary glands to the *Dermacentor andersoni* larvae.

Later, in 1979, Allen et al. carried out another interesting trial: after having proved that the blind guts, the reproductive system, etc., of the *Dermacentor andersoni* which had not been in contact with a resistant host caused no reaction when inoculated into guinea pigs, they started an experiment with cattle ticks which had been feeding for 5 days on resistant guinea pigs. After dissecting the cattle ticks, they prepared two antigens: one to be called A, composed of an extract of reproductive organs and guts; and another to be called B, which was a maceration from all the parasites' inner organs.

Eighteen guinea pigs were selected for the trial and divided into three groups of six each. Those from the first group were inoculated with 1.2 mg of antigen A (2 doses at 14-day intervals); those from the second group were likewise inoculated with antigen B, whereas the control, the third group, was only inoculated with the adjuvant. Twenty-eight days after the experiment had begun, all the guinea pigs were infested with four *Dermacentor andersoni* females and four males, which 2 weeks later were withdrawn.

Both parasites and oviposition were collected and weighed, and 35 days after the ticks had been collected, the eclosion of eggs was studied. The results, in brief, are those detailed in Table 7.2.

As can be observed, the results are significant enough to confirm the hypothesis of the authors, who after their first success, began to work with cattle by inoculating steers with 67 mg of antigen A administered intramuscularly, at 16-day intervals, adding a further third inoculation (containing no adjuvant whatsoever) on the 25th day.

In turn, the steers were infested with 10 parasites each on day 28 and the parameters detailed in Table 7.3 were also studied in depth.

As can be seen, except in the number of parasites collected (which show no significant differences), the results are in general undoubtedly important, and could open the way to future investigations.

Table 7.2. Mean weight of engorged females, weight of eggs laid and average larvae per engorged female

	Mean weight in mg of ticks (female)	Weight of spawning in mg in relation to female	Average larvae per female
Controls	1483.84	733.34	10.725
Immunized with A antigen	629.51	126.95	0
Immunized with B antigen	165.2	0	0

Table 7.3. Comparison between ticks collected from vaccinated animals and controls

	Total ticks recovered (average)	Total weight of ticks (average)	No. of female that spawned	Total No. of eggs ($\times 10^3$)	Total No. of larvae ($\times 10^3$)
Animals immunized	97	4.35	15	27.6	23.2
Controls	107	24.44	50	136.8	115.5

Apart from the Australian and Canadian trials already described, we have heard of no other investigation on this subject, but as an anecdote, we can quote a very elemental trial carried out in Argentina (Menard et al. 1971 unpublished), where on oily vaccine was prepared with an homogenate of *Boophilus microplus* larvae. Five cm^3 of this vaccine were administered subcutaneously to steers which had had no contact with the parasite, leaving three others as controls. Twenty-one days later, each animal was infested with 10,000 larvae, aged 3 weeks. Yet no significant results were obtained after studying the number of engorged females collected, the oviposition and later eclosion.

D Grooming

This is the last of the factors we shall analyze. The word *grooming* refers to cattle licking themselves to get clean or, in this particular case, to detach the ectoparasites. The problem of host resistance to *Boophilus microplus* has been studied in detail by many investigators, among them, Kelley (1943), Lombardero et al. (1970), Riek (1956) and Villares (1941). As a result of these studies, these investigators observed that those animals which had actively scratched and groomed themselves after infestation underwent lower parasitic loads than those which remained quiet during infestation.

It is really very difficult to know whether this fact is due to an individual characteristic or is caused by either an allergic or an immune reaction.

Snowball (1956) worked with a cattle herd which he divided into two lots: he restrained one of them from grooming by means of an appropriate harness, while the other was left unrestrained. After infesting both lots and later collecting the detached engorged females − once the parasitic cycle was completed − Snowball obtained the following results: in those animals from the first lot, which had been restrained from grooming, 33% of larvae fulfilled their cycle until becoming engorged females (extreme data between 5 and 69%); in the controls, that is those under no restraint, the percentage was of 91% (between 0.1 and 32%). Taking into account this information, the author came to the following conclusion: both grooming and partially scratching (which causes death to the parasites by mechanical action) constitute an important system with respect to the control of parasitic populations.

The work of Bennett (1969) infesting both resistant and susceptible animals, and restraining them from grooming, demonstrates that the collection of engorged females from resistant animals notably increases from the 3rd to the 5th week within the restraining period, later decreasing abruptly, which would indicate, according to this Australian author, that self-grooming is an important factor in the host resistance to the ixodide; but that the restraint from grooming also increases the power of a later immunity response.

The Inheritance of Resistance

Once the possibilities of taking advantage of the host's resistance to the parasite as a control factor have been analyzed and in spite of its limitations, the question arises as to whether it is hereditary and, if so, in what proportion.

According to Hewetson (1976), the mechanism by which cattle acquire resistance to the common cattle tick can be inherited to such a degree that it justifies a selective process in order to achieve a high level of parasite control. He also writes that several other studies have been carried out on this subject, the results in general ranging from 40 to 50% inheritability in European cattle.

Seifert (1971) considers that the selection of cattle with this trait in European cattle is very difficult due to its low inheritability; and Wharton et al. in 1970 — a period of hard times in Australia due to the appearance of *Boophilus microplus* strains resistant to nearly all the commercial ixodicides — based the highest possibilities of control upon the selection of cattle resistant to ixodide.

Conclusions

On evaluating the varied and copious information available on this subject and taking into account our experience in Argentina, we can draw some conclusions, the most important of which are:

A) Resistance to *Boophilus microplus* infestation in Indian cattle breeds is generally accepted.
B) Bovines of European breeds (except Jersey) are a quite different case; their resistance to the parasites is an unusual characteristic.
C) In spite of the fact that either grooming or hypersensitivity may play an important role in resistance, everything leads us to suppose that the fundamental mechanism is of an immunological nature; but whether it is of a humoral or cellular kind, or more probably a combination of both, is still under dispute.

D) The level of resistance inheritance in European breeds is quite low, thus rendering the process of selection extremely difficult.

In brief, although selection of cattle resistant to *Boophilus microplus* may be considered, at least in theory, as a means of controlling cattle tick, we consider that the existing ixodicides, extremely safe and effective as they are, together with both the better technology in use and possibly with control by means of either working or chemical systems during the parasite free-living cycle, are all much more effective methods of controlling or even eradicating *Boophilus microplus* than the selection of cattle resistant to the cattle tick itself; since the relative character of this resistance, as well as its low inheritance rate in European cattle, may lead, at best, to zootechnical concessions that would result in the decrease of production, a fact which would lead to a worse economic situation than that caused by the present systems used to fight cattle tick.

References

Allen JR (1973) Tick resistance: basophils in skin reactions of resistant guinea pigs. Int J Parasitol 3:195 – 200

Allen JR, Humphreys SJ (1979) Immunization of guinea pigs against ticks. Nature (London) 280:491 – 493

Allen JR, Wikel SK (1978) Acquisition of tick resistance in guinea pigs. Tick-borne diseases and their vectors 75 – 76. Univ Edinburgh

Allen JR, Khalil HM, Wikel SK (1979) Langerhans cell trap tick salivary gland antigens in tick-resistant guinea pigs. J Immunol 122:563 – 565

Bennett GF (1969) *Boophilus microplus (Acarina: Ixodidae)* Experimental infestations on cattle restrained from grooming. Exp Parasitol 26:323 – 328

Campbell RSF (1978) The use of resistant cattle in the control of tick and vector-borne diseases. Tick-borne diseases and their vectors. Univ Edinburgh, pp 251 – 257

Findlay JD, McEwan Jenkinson D (1960) The morphology of bovine sweat glands and the effect of heat on the sweat glands of the Ayrshire calf. J Agric Sci 55 (2):247 – 250

Hewetson RW (1968a) Resistance of cattle to cattle tick *Boophilus microplus.* I) The inheritance of resistance to experimental infestations. Aust J Agric Res 19:323 – 333

Hewetson RW (1968b) Resistance of cattle to cattle tick *Boophilus microplus.* II) The inheritance of resistance to experimental infestations. Aust J Agric Res 19:497 – 505

Hewetson RW (1971) Resistance by cattle to cattle tick *Boophilus microplus.* III) The development of resistance to experimental infestation by purebred Sahiwasl and Australian Illawarre Shorthorn cattle. Aust J Agric Res 22:331 – 342

Hewetson RW (1972) The inheritance of resistance by cattle to cattle tick. Aust Vet J 48:299 – 303

Hewetson RW (1976) Selection of cattle for resistance against *Boophilus microplus.* Proceedings of tick-borne diseases and their vectors. Centre Trop Vet Med, Univ Edinburgh, pp 258 – 261

Ivancovich JC (1970) Grado de parasitación por *Boophilus microplus* en ganado bovino de origen europeo. V. Jornadas Int C Vet La Plata Tomo II:41 – 48

Johnston TH, Bancroft MJ (1918a) A tick-resistant condition in cattle. Proc Soc Queensl 30:219 – 317

Johnston TH, Bancroft MJ (1918b) A tick-resistant condition in cattle. Proc R Soc Queensl 41:121 – 132

Kelley RB (1943) Zebu-cross cattle in Northern Australia. An ecological experiment. Counc Sci Indust Res Bull 172, Melbourne, Aust

Kemp DH (1978) In vitro culture of *Boophilus microplus* in relation to host resistance and tick feeding. Tick-borne diseases and their vectors. Univ Edinburgh, pp 95 – 99

Kemp DH, Koudstaal D, Roberts JA, Kerr JD (1976) *Boophilus microplus*: the effect of host resistance on larval attachments and growth. Parasitology 73:123 – 136

Koudstaal D, Kemp DH, Kerr JD (1978) *Boophilus microplus*: rejection of larvae from British breed cattle. Parasitology 76:379 – 386

Legg J (1930) Some observations on the life history of the cattle tick (*Boophilus australis*). Proc R Soc Queensl 41:121 – 132

Lombardero OJ, Schiffo HP (1968) Diferencias tróficas y biológicas en *Boophilus microplus* de cebú y su derivado media sangre. Rev Fac C Vet La Plata 22:167 – 173

Lombardero OJ, Schiffo HP, Pastori RB (1970) El *Boophilus microplus* como parásito del cebú. Nuevas observaciones. V. Jornadas Int C Vet La Plata Tomo II:49 – 54

Nicolaides N, Fu Hwei C, Rice GR (1968) The skin surface lipids of man compared with those of 18 species of animals. J Invest Dermatol 51:83 – 89

Riek RF (1956) Factors influencing the susceptibility of cattle to tick infestation. Aust Vet J 32:204 – 209

Riek RF (1962) Studies on the reactions of animals to infestation with ticks. VI Resistance of cattle to infestation with the tick *Boophilus microplus*. Can Aust J Agric Res 13:532 – 550

Roberts JA (1968) Acquisition by the host of resistance to the cattle tick *Boophilus microplus* (Canestrini). J Parasitol 54:657 – 662

Roberts JA, Kerr JD (1976) *Boophilus microplus*: passive transfer of resistance in cattle. J Parasitol 62 (3):485 – 489

Schleger AV, Lincoln DT, McKenna RV, Kemp DH, Roberts JA (1976) *Boophilus microplus*: cellular responses to larval attachment and their relationship to host resistance. Aust J Biol Sci 29:499 – 512

Seifert GW (1971) Variations between and within breeds of cattle in resistance to field infestations of the cattle tick (*Boophilus microplus*). Aust J Agric Res 22:159 – 168

Smith ME, McEwan Jenkinson D (1975) The effect of age, sex and season on sebum output of Ayrshire calves. J Agric Sci 84:57 – 60

Smith ME, Ahmed SU (1976) The lipid composition of cattle sebaceous glands: a comparison with skin surface lipid. Res Vet Sci 21:250 – 252

Smith ME, Noble RC, McEwan Jenkinson D (1975) The effect of environment on sebum output and composition in cattle. Res Vet Sci 19:253 – 258

Snowball GJ (1956) The effect of self-licking by cattle on infestations of cattle tick *Boophilus microplus* (Canestrini). Aust J Agric Res 7:227 – 232

Sutherst RW, Wharton RH, Cook IM, Sutherland ID, Bourne AS (1979) Long-term population studies on the cattle tick (*Boophilus microplus*) on untreated cattle selected for different levels of tick resistance. Aust J Agric Res 30:353 – 368

Tatchell RJ, Bennett GF (1969) *Boophilus microplus*: antihistaminic and tranquilizing drugs and cattle resistance. Exp Parasitol 26:369 – 377

Utech KBW, Wharton RH, Kerr JD (1978) Resistance to *Boophilus microplus* (Canestrini) in different breeds of cattle. Aust J Agric Res 29:885 – 895

Villares JB (1941) Climatología zootechnica. III Contribucao ao estudo da resistencia e suspectabilidade genetica dos bovinos ao *Boophilus microplus*. Bol Indust Anim 4:60 – 80

Wagland BM (1979) Host resistance to cattle tick *Boophilus microplus* in Brahman (*Bos indicus*) cattle. IV Ages of ticks rejected. Aust J Agric Res 30:211 – 218

Wharton RH (1975) Resistant cattle for biological control of cattle tick. Anim Quarantine 4 (2):13 – 17

Wharton RH, Norris KR (1980) Control of parasitic arthropods. Vet Parasitol 6:135 – 164

Wharton RH, Utech KBW, Turner HG (1970) Resistance to the cattle tick *Boophilus microplus* in a herd of Australian Illawarra Shorthorn cattle: its assessment and heritability. Aust J Agric Res 21:163 – 181

Wikel SK (1979) Acquired resistance to ticks. Expression of resistance by C4-deficient guinea pigs. Am J Trop Med Hyg 28 (3):586 – 590

Wikel SK, Allen JR (1976a) Acquired resistance to ticks. I. Passive transfer of resistance. Immunology 30:311 – 316

Wikel SK, Allen JR (1976b) Acquired resistance to ticks. II. Effects of cyclophosphamide on resistance. Immunology 30:479 – 484

Wikel SK, Allen JR (1978) Characterization of the immune response in tick-resistant guinea pigs. Tick-borne diseases and their vectors. Univ Edinburgh, pp 77 – 78

Wikel SK, Graham JE, Allen JR (1978) Acquired resistance to ticks. IV Skin reactivity and in vitro lymphocyte responsiveness to salivary gland antigen. Immunology 34:257 – 263

Wilkinson PR (1955) Observation on infestations of undipped cattle of British breeds with the cattle tick *Boophilus microplus* (Canestrini). Aust J Agric Res 6:655 – 665

Wilkinson PR (1962) Selection of cattle tick for resistance and the effect of herds of different susceptibility on *Boophilus* populations. Aust J Agric Res 13 (6):655 – 665

Willadsen P, Williams PG (1976) Isolation and partial characterization of an antigen from the cattle tick *Boophilus microplus*. Immunochemistry 13:591 – 597

Willadsen P, Williams PG, Roberts JA, Kerr JD (1978) Responses of cattle to allergens from *Boophilus microplus*. Int J Parasitol 8:89 – 95

Willadsen P, Wood GM, Riding GA (1979) The relation between skin histamine concentration, histamine sensitivity and the resistance of cattle to the tick *Boophilus microplus*. Z Parasitenkd 59:87 – 93

Yang SH (1951) Histochemical studies on bovine sweat glands. J Agric Sci 42:155 – 158

VIII Resistance

Resistance is the development of a condition, in a population of ticks and other arthropods, that permits toleration of toxic doses otherwise lethal to members of a normal population of the same species. The concept of resistance is described by FAO in this simple and concise definition which, in essence, is the capacity that living beings have to adapt some of their many variables to a change of medium or environment, thus ensuring the survival of the species.

Boophilus microplus, the common cattle tick, is not an exception to this rule. From studies carried out by Brown (1967) and evidence obtained from various investigators, it was concluded that resistance to a certain active compound or a group of active compounds is a hereditary characteristic of *Boophilus microplus*, developed from the selective pressure of tickicides over "pre-existing" resistant mutants whose proportion increases as the strain comes into contact with the ixodicide. Therefore induction (the more defective the use of the acaricide, the more important or effective will induction be) plays an important part, as once the resistance is established it acts as an element of selection.

This strengthens the stated theory and makes it highly improbable that resistant genes appear as a result of acaricide treatments.

Clinical Symptoms

Diagnosis

The appearance of resistance occurs sporadically; it must be duly confirmed and requires a careful evaluation of the following factors:

A Anamnesis: two key points are stressed.
A1 Handling: in most cases the possible findings of resistance are a result of incorrect usage of the ixodicide; for example: insufficient immersion time, incomplete immersion of the head, reduced concentration of the active ingredient, etc.
A2 Type of active ingredient: it is basic to enquire for how long the active ingredient has been used in the establishment, and, if another product was used

previously, if it belonged to the same chemical group, for then it could be thought of as cross-resistance or partially "inherited".

B Check the concentration of active ingredient: this step is indispensable. If it is not confirmed whether or not the ixodicide is at the correct level, the possibility of resistance cannot be taken seriously.

C Carry out the clinical trial: despite the development of several techniques for studying susceptibility to ixodicides using larvae or engorged females, which very often serve as an important guide in diagnosis, the decisive trial consists in carrying out the treatment with a correct handling and adequate concentration of the active ingredient, to animals infested with the suspect strain.

Symptoms

At first the most classic symptom of resistance of ticks to an ixodicide is the survival of some engorged nymphs that develop normally until, 10 – 14 days later, they become engorged females. In more advanced cases semi-engorged females are found and as the process continues undistended females and males are seen, that normally are very sensitive to tickicides. In extreme cases, not described in Argentina, we have seen parasitic descriptions that are practically unmodified by treatment.

In the first stages, when the symptoms are not yet clear, anamnesis and checking of the correct active ingredient are sufficient for a presumptive diagnosis which will then be confirmed by the clinical trial.

Our experience tells us that no morphological characteristics can be appreciated that distinguish a resistant strain from a susceptible one, no suggestive detail having been noted. Studying oviposition of females belonging to three resistant strains, Bennett (1975), in Canada, observed that it was not significantly different to the strains used as controls, adding that the acquisition of resistance does not seem to affect other biological parameters. It is understood that in this trial the ticks used were unaffected by tickicides.

Mechanisms

The mechanisms by which *Boophilus microplus* developed resistance in different countries can be classified in three main groups:

A Alteration of the acaricide properties in the site of action.

B Alteration of the rate of metabolism of the ixodicide.

C Alteration in its rate of vehiculization. Although this mechanism has not yet been demonstrated in the few cases studied, we consider it potentially of great importance.

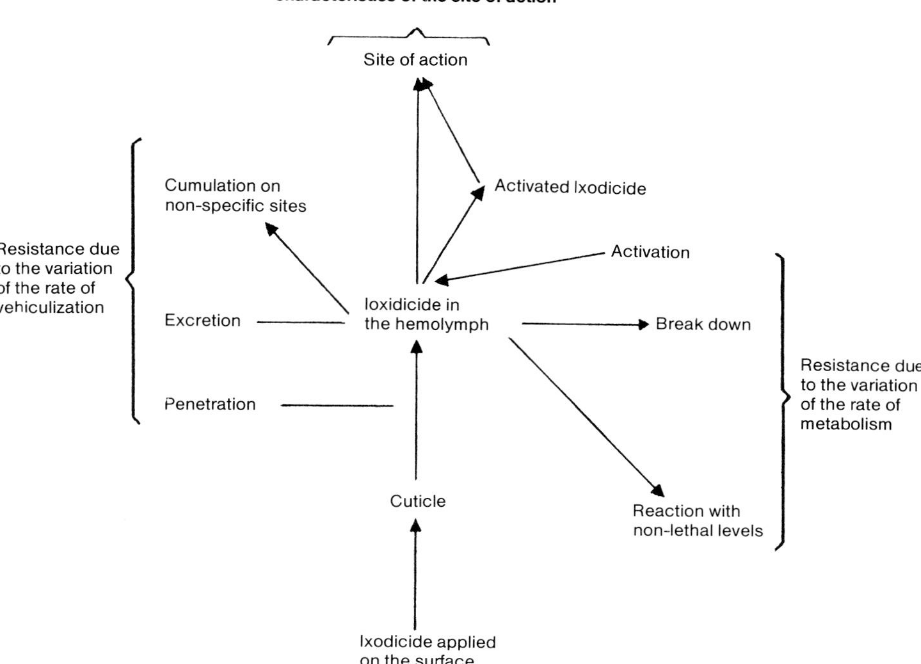

Fig. 8.1. Biochemical mechanisms that cause parasite resistance. Wilson (1978)

In the graph of Fig. 8.1, published by Wilson in 1978, there is an accurate and simple description of the biochemical mechanisms that cause tick resistance to ixodicides.

The arsenicals were the first tickicides used this century, later replaced by the chlorides; these were followed by phosphorates and carbamates, and at present therapeutics are based on trizapentadiene, formamidines, cycloamidines and synthetic pyrethroids. We will analyze the first three groups, in which the development of resistance was evident and very widespread, but we will not describe its mode of action, as it has been treated in the chapter on therapeutics.

Arsenicals

Mechanism of Resistance

Studies carried out by Whitehead (1958) in South Africa, showed that the level of free sulfhydrylic groups was higher in resistant strains than in susceptible ones.

This excess of sulfhydrylic groups is capable of annulling the arsenic, thus reducing the acaricide concentration in the site of action.

Background

The salts most commonly used as tickicides were the arsenites, particularly sodium methararsenite, obtained from the reaction between arsenic trioxide (AS_2O_3) and caustic soda (OHNa), which replaced the simple preparations such as solutions of arsenic trioxide (AS_2O_3).

In an important contribution that gathers trials carried out in Argentina between 1940 and 1949, Ault (1948) reports that at that time there were already problems of resistance (though not so serious or extensive as they were thought to be) and that the difficulties in cleaning the animals (those that were transported to areas free of ticks) at the normal concentrations of AS_2O_3 0.2%, increased with time.

In South Africa, Du Toit et al. (1941) described cases of resistance in *Boophilus decoloratus*.

In Australia, where arsenic had been used since 1895, the first signs of resistance appeared in 1937 and gradually extended until the fight against *Boophilus microplus* became very difficult.

Chlorides

Mechanism of Resistance

The mechanism of the toxic effect of chlorides on arthropods was practically unknown, and if symptoms ranging from an inhibition of mesoinositol at cell level to a simple accumulation of particles in the nervous cell membrane were mentioned, it was no more than theoretical speculations that were not possible to prove scientifically.

For this reason the mechanism of resistance to organochloride compounds is unknown, even if its neurotoxic function is recognized.

Background

After arsenical compounds had been used for several decades, as from 1946 they began being replaced by chlorides, of which the HCH (hexachlorocyclohexane) isomer was the most used in Argentina. Although on a smaller scale, other compounds used were toxaphene, DDT (dichlorodiphenyltrichloroethane) and dieldrin.

The first clinical symptoms of resistance were observed in 1952, and reported already in 1953 by Boero. From that time on, resistance became more and more widespread in Northeast Argentina, until it seriously affected tick control, endangering eradication plans.

One fact can justify and explain the rapid appearance of HCH, or rather, rapid selection of resistant strains. It is the greater speed of degradation of gamma-isomer compared to the other isomers (alpha, beta, delta and epsilon), which made the concentration of active ingredient in plunge dips decrease rapidly. As at the time the analytical technique was based on titration of HCH total and then its percentual distribution according to the original theoretic content of isomers, the true gamma-isomer of HCH was always below the stated one. This interesting and useful work by Allan was published in 1955 but did not get the publicity deserved in Argentina, where it was not analyzed until several years later.

Regarding other countries we can say that both in South Africa and Australia resistance to HCH appeared in 1952, between a year and a half and two years after it had started to be used, which makes us suppose that there was a problem similar to that described in the previous paragraph. Regarding DDT, resistance was demonstrated in Australia in 1955, while in South Africa it occurred in 1956. Here a curious thing happened: in Australia the DDT-resistant strains were susceptible to dieldrin and gamma-isomer of HCH, while in South Africa they were also resistant to arsenic, HCH, and showed a crossed resistance to pyrethrum (they quote Harrison et al. 1973).

We must note that in Argentina strains resistant to either DDT or dieldrin were never isolated, a fact of great importance in view of the possible use of synthetic pyrethroids, for, as we will see later, there are apparently crossed resistance phenomena between these active ingredients.

Finally, to complete the information on resistance to organochlorides in other countries, we can add to those already mentioned findings in India (Chaudhur and Naithani 1964), Madagascar (Vilenberg 1963) and in the French Antilles (Morel 1967).

Organophosphorous Compounds and Carbamates

Mechanisms of Resistance

In the mid 1950's organophosphorates began to be used as the only and therefore logical replacement of organochlorides. Their mechanism of resistance is, as is to be supposed, closely related to the effect explained previously.

The use of this group of chemicals became extensive during the 1960's, and already in 1963 in Rockhampton, Australia, the first resistant strain, named Ridgelands, was isolated. This strain surpassed the ixodicide effect of dioxathion and diazinon, and Lee and Betham (1966) showed that its mechanism of action was

Table 8.1. Cholinesterase activity of resist. Cholinesterase activity of resistant and susceptible strains, according to Lee and Batham (1966)

Species and Strain	No. of homogenates	Cholinesterase activity (moles/g/h)	Standard deviation
Boophilus microplus Susceptible strain	10	370 ±	23
Boophilus microplus Resistant strain	20	75 ±	13

based on the fact that the strain had at least one cholinesterase reacting more slowly than that of a strain susceptible to inhibition by an organophosphorate; in other words, that the cholinesterase activity was much less in a resistant strain than in a susceptible one. To prove this, they prepared homogenates of larvae ($50\,\text{mg ml}^{-1}$) of 2 to 3 weeks of age, mixed at a rate of 0.5 ml homogenate, 0.5 ml buffer (pH 7.2) and 1 ml acetylcholine perchlorate ($8 \times 10^{-3}\,\text{M}$), incubated at $37\,^{\circ}\text{C}$ and determining residual acetylcholine with the method of Hestrin.

The cholinesterase activity measured in micromoles of hydrolysed acetylcholine/g of ticks/time, in the case described by these authors, showed the differences illustrated in Table 8.1.

In 1966 in Queensland, Australia, another strain, named Biarra, was isolated (Roulston et al. 1967); it presented a very broad spectrum of resistance, including all the tickicides registered in the country up to the moment, they were:

Organophosphorate	*Factor of resistance*
Carbofenothion	39.3
Ciodrin	6.3
Coumaphos	24.8
Diazinon	38
Dioxathion	16.2
Ethion	18.6
Carbaryl (Carbamate)	7.5

These figures clearly reflect the alarming level of resistance which at that time endangered the programme of ixodide control in the country (Fig. 8.2).

The mechanism of resistance in the Biarra strain had basically the same characteristics as the Ridgelands, i.e., it was based on a modified cholinergic system.

A third strain, called Mackay, was isolated later in a town of that name, almost 1000 km North of Biarra in Brisbane Valley, and its mechanism of resistance was studied in detail by Roulston and Wharton (1969). Its main character-

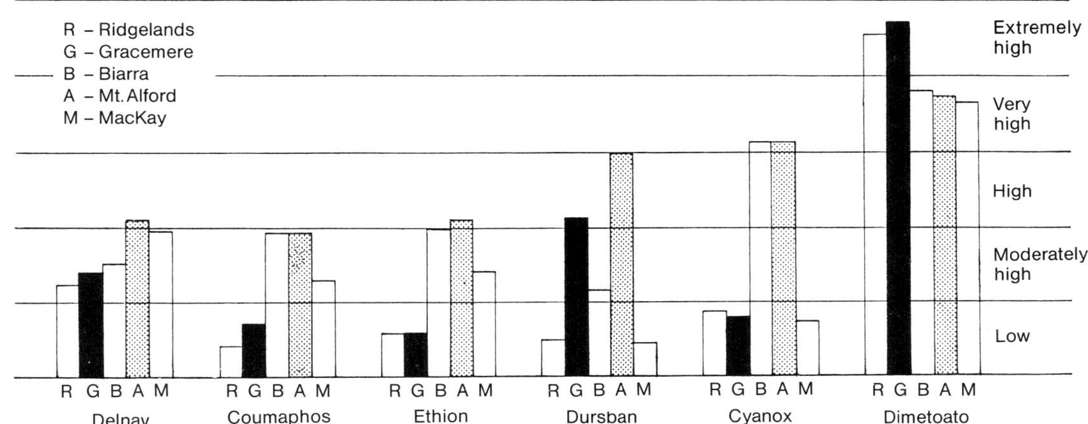

Fig. 8.2. Factors of resistance in larvae of resistant strains of *Boophilus microplus*. Stone and Haydock Technique (1972). Published in Stone (1972)

istics were the intermediate resistance between Ridgelands and Biarra strains to ethion and coumaphos and the sensitivity of acetylcholinesterase to other organophosphorous compounds, all of which led to the idea of different mechanisms of resistance. This was confirmed when trials were carried out with coumaphos marked with ^{32}P and the metabolism of this tickicide was compared in Yeerongpilly (susceptible), Biarra and Mackay strains, observing in this latter a process of detoxification due to a much faster degradation of the acaricide than in the other strains studied, which gave as a result a reduced concentration and persistance of the active ingredient in the site of action.

Later, new strains were isolated, always in Queensland (Mt. Alford, Gracemere and Silkwood), studied by Schnitzerling et al. (1974) where, in a general way, the two mechanisms described previously were observed. For example, the results of laboratory trials carried out with Mt. Alford strain are given in a publication by O'Sullivan and Green (1971); they correspond to the factor of resistance of Biarra strain:

Dursban	100	(3.8)
Diazinon	311	(64.0)
Coumaphos	23	(18.0)
Carbaryl	10	(16.0)
Ethion	16	(18.0)
Bromophos ethyl	2.2	(3.7)

The same authors make reference to other strains, such as Kenilworth and Pimpana, which can be considered within the same group. For carbamates, everything indicates that the mechanism of resistance developed by *Boophilus microplus* is similar to that corresponding to organophosphorates.

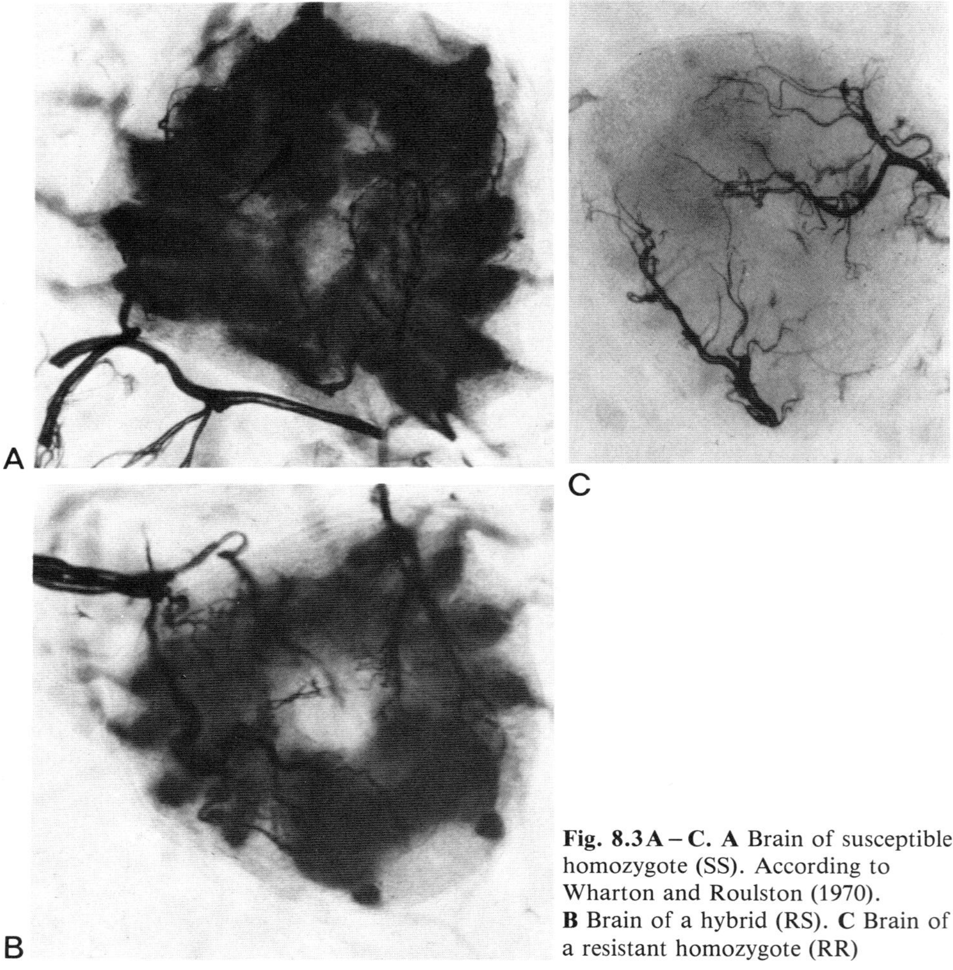

Fig. 8.3A – C. A Brain of susceptible homozygote (SS). According to Wharton and Roulston (1970). **B** Brain of a hybrid (RS). **C** Brain of a resistant homozygote (RR)

We have already quoted the work done by Lee and Batham, where a lesser cholinesterase activity in phosphorous-resistant strains was shown, by means of trials carried out with homogenate of larvae. This led to the study of cholinesterase activity in the central nervous system in susceptible and resistant individuals. Wharton and Roulston (1970) quote, referring to a publication by Stone (1968a), that processing brains of homozygous resistant (RR) individuals, a cholinesterase activity of approximately 12% was observed, of which they showed the brains of susceptible homozygotes (SS), while their hybrids (RS) had around 78% of the activity of their susceptible parents (Fig. 8.3, A, B and C).

Background

The information obtained for studying the different mechanisms of resistance to organophosphorous compounds, gave us a detailed idea of what happened in Australia, where the problem was most extensive and where it was studied in depth.

In Argentina, the first reference was made in 1964. At that time the strain named Las Guerisas was isolated in the province of Entre Ríos. Fluck and Rufenacht published the finding in 1969. This strain seemed to correspond with the Ridgelands type resistant to diazinon and delnav, and was apparently eradicated using ethion and/or coumaphos, for the area where it was found was declared free in 1968 – 1969 (Panel de Garrapata 1972).

In 1969, Grillo Torrado and Gutierrez described the strain named Goya, which showed a high resistance to coumaphos when studies were carried out on larvae using a technique devised by the authors. We have also worked on this strain (Nuñez et al. 1972) finding a moderate factor of resistance to coumaphos ($\times 5$) and a not very significant one to ethion ($\times 2.3$), a fact that coincides with the clinical trials carried out by government organizations, who demonstrated the efficacy of the ixodicide under field conditions.

It must be stressed that the Goya area, in the Southwest of the province of Corrientes, Argentina, is still within the campaign area, therefore ixodicides are used intensively. It is easy to realize that as 12 years ago there was such a high index of resistance, and taking into account that one of the fundamental genetic characteristics of resistance of *Boophilus microplus* to organophosphorates is the incomplete dominance of this characteristic, the present description of the area should show a generalized resistance, which is not the case (Oria 1979 personal communication).

During the last few years, particularly 1979 – 1980, studies of resistance to organophosphorous products have been intensified, due to the increasing number of problems in the campaign area.

In 1980 (personal communication), Oria and colleagues isolated the María Isabel strain, with clinical symptoms of resistance and, on being studied in the laboratories of SENASA (Animal Health Service of the Ministry of Agriculture) it gave a factor of resistance to ethion of 9. This figure coincides with our study carried out simultaneously, and which is the first description in the country of resistance to ethion. They also gave the following factors of resistance to other ixodicide compounds:

Coumaphos	6.8
Supona	7.0
Bromophos ethyl	5.3
Amitraz	0.63

Information provided by Caracostantogolo of SELAB (Ministry of Agriculture), indicates that in a census carried out in Mercedes, province of Corrientes,

from 14 samples, 4 were found to be resistant, 3 of which to more than 2 tickicides:

Strain A: resistant to coumaphos, supona, ethion, bromophos ethyl and chlor-
pyriphos.
Strain B: resistant to coumaphos, ethion, and chlorpyriphos.
Strain C: resistant to coumaphos, supona and ethion.

The remaining strain was resistant to supona and ethion.

All this refers mainly to laboratory trials.

In another census carried out in Goya, of 13 strains studied, there were no cases of resistance; on the other hand, in Curuzú Cuatiá (Corrientes, Argentina), 4 strains that presented clinical symptoms of resistance were isolated, among them María Isabel which we have already mentioned. The other 3 showed moderate resistance to coumaphos, supona and ethion, and only in one case to bromophos ethyl. Finally, in the province of Santa Fe, 14 samples were studied without detecting any signs of resistance.

Similar results were obtained from two samples sent from the province of Santiago del Estero.

In 1968 Shaw et al. described strains of Ridgelands type in other South American countries, namely, in Brazil and Venezuela.

In 1970 the strain named Guaimarito was isolated in Venezuela; its level of resistance is placed between Ridgelands and Biarra, but with biochemical characteristics similar to the former, for which it is usually classified as the Super Ridgelands type strain. Some of its factors of resistance measured in larvae using the Shaw technique are: delnav 60, ethion 21, coumaphos 10 and supona 6.

Already in all these South American countries organophosphorous products are used in very low proportions, as in Northern Uruguay, where resistance to this group of active compounds is spreading rapidly (Cravino and Genovese 1980 personal communication).

Finally, in Africa, Jones Davies (1972), in a complete summary on *B. decoloratus* resistance, refers to the Berlin strain, giving the following coefficients of resistance to Carbaryl and various other organophosphorates.

Carbaryl	75
Trithion	51
Dioxathion (delnav)	16
Diazinon	40
Ethion	3
Chlorfenvinphos (supona)	2
Coumaphos	2

In Zimbabwe, Matthewson et al. (1976) isolated 8 strains of *B. decoloratus*, with variable levels of resistance to several organophosphorates and in 4 of them to carbaryl as well.

Biochemical studies on one of these strains (Alicedale West Farm) indicated a level of resistance between Berlin strain of *B. decoloratus* and Biarra of *Boophilus microplus* (Matthewson et al. 1976).

Synthetic Pyrethroids

This is one of the groups that provides the greatest number of compounds of considerable efficacy as tickicides, the commonest and most studied being permethrin, cypermethrin, cypothrin and especially deltamethrin. Up to the present they have not been used on a large scale under field conditions, but already in 1977 Nolan et al. in Australia noticed a cross-resistance between DDT-resistant strain and this group of compounds, which was demonstrated in the laboratory and in the field.

This fact might not be very important in Argentina where the use of DDT was of little consequence, therefore there are no real possibilities of having selected a strain resistant to this organochloride.

Evolution of Resistance After Relaxation of Chemical Pressure

Once resistance is established and a chemical product or group of chemical products made useless, it is natural to ask if this fact is temporary or if it remains latent and becomes evident again with moderate stimulation.

According to our experience, arsenical resistant strains without contact with arsenicals for several years, when treated with this active ingredient, quickly reached their old level of resistance; this fact was also described in Australia by Roulston and colleagues in 1971 with Biarra and Mackay strains (quoted by Stone 1972).

Also in Australia, Stone gives the following information:

DDT-resistant strain: after relaxation of chemical pressure, the level of resistant homozygotes falls from almost 100% to 55% in 13 generations, and stays that way for another 18 generations (Stone 1962).

Strains resistant to HCH-Dieldrin: in this case, from 50 – 60% resistant individuals, a percentage near 0 and 24 generations drops off (Stone 1962a).

Strains resistant to organophosphorates: Five generations after pressure by an acaricide was halted, resistance remained steady and, in some cases, increased slightly (Stone 1968a).

These are *changes in the proportion of resistant genotypes,* but do not affect their *level of resistance,* which remains unchanged even in the absence of ixodicide treatment.

All this leads to the conclusion that there is little possibility of re-using an active principle to which *Boophilus microplus* develops resistant strains at any time.

Genetics of Resistance

When trying to study genetic aspects in *Boophilus microplus*, there are many important problems, for, added to the prolonged life cycle are problems of feeding, crossing, complementary biochemical studies, etc., for which appropriate installations and expensive special laboratory material are necessary.

Only in Australia, where the problems of resistance to tickicides in the common cattle tick reached uncommon levels was work of an excellent scientific level done and, with due precautions, reference can be made to projects in Africa and South America, where the clinical and biochemical frames of resistance on the whole follow a similar development.

The trials by Stone (1972) and Oliver (1967) tell us that the diploid number of chromosomes in *Argasidae* is $12-32$ and in *Ixodidae* $21-28$. In *Boophilus microplus* (as in *Ripicephalus sanguineus*) the genetic formula is:

$$2n = 21 \text{ (male)} = 22 \text{ (female)}$$

and sex is determined by the mechanism of XX-XO.

It is also believed that crossing-over takes place, thus permitting the free combination of resistant genes.

The sex chromosomes are very large, several times larger than autosomes, which do not differ much from each other.

The common methods of reciprocal individual crossing of resistant and susceptible homozygotes are used, carrying out the corresponding backcross of heterozygotes obtained by one or both parents according to Stone (1972). This same author maintains that in all the cases studied, resistance is due to only one pair of autosomic genes or to a closely related group of autosomic genes, which in this case would occupy the same locus.

As proof of the monofactorial inheritance of organophosphorous, we reproduced the graph (Fig. 8.4) which summarizes studies done by Stone. In these studies, having evaluated the susceptibility of the parents (S and R), the scientist carries out crossing (F_1) and back-crossing ($F_1 \times S$ and $F_1 \times R$).

The evaluations of sensibility are carried out using the techniques of Stone and Haydock on a group of 120 to 240, $7-11$ days old larvae, using formothion as an acaricide.

By following the theoretical lines of the monofactorial inheritance in both back-crosses, it can be seen that the results of the experimental trials have regular intervals (mortality – acaricide concentrations).

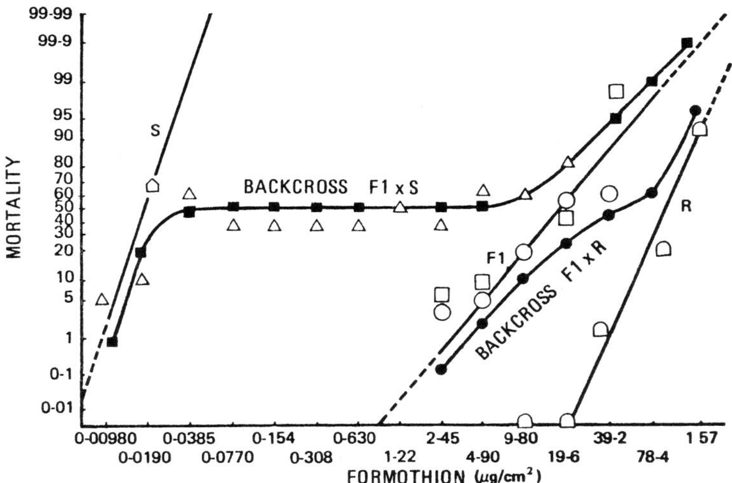

Fig. 8.4. Mortality of *Boophilus microplus* larvae. Stone (1968)

Finally, once again Stone (Stone and Haydock 1962, Stone 1962b) — whose works were complemented by Brown (1967) — came to the conclusion that resistance to DDT is incompletely recessive; that resistance to dieldrin has complete dominance; and resulting from organophosphorous compounds (he studied the Ridgelands strain) has characteristics of incomplete dominance, a fact that was later confirmed in the Biarra strain (Wharton and Roulston 1970).

More recently, Stone et al. (1976b), studied Ridgelands Biarra and Mackay strains and compared them to a susceptible strain, from which they provided interesting data on different biochemical aspects of resistant genetics, for example, the reduced cholinesterase activity on brains of adult stages of the Biarra and Mackay strains; less sensibility to cerebral acetylcholinesterase inhibiting organophosphorates in specimens of the three strains; and the increased detoxification in larvae and adults of the Mackay strain. By means of microspectrophotometric measurements of the acetylcholinesterase activity in brain histochemical preparations, they also proved that the hybrids have an approximate level of activity which is intermediate between that of the parents, and confirmed that the lower enzymatic activity of the Biarra strain is controlled by a closely linked pair of autosomal genes.

In Argentina, Grillo Torrado et al. (1972) studied the B strain of *Boophilus microplus* (isolated in the province of Corrientes) crossing it with A susceptible and, measuring the sensibility of hybrids, proved that resistance is not related to sex nor does it depend on extrachromosomic factors; that the degree of dominance was 0.48 and that resistance in the B strain could depend upon only one pair of autosomic genes or perhaps a group of closely related genes.

Also in Buenos Aires, Reich et al. (1978) studied the differences between the cholinergic systems of a sensible strain and another resistant one, establishing maximum activity at pH 6, 7 – 7.8. Centrifugation at 27,000 g caused separation of the two fractions, one of them soluble; in the resistant strain it showed a marked reticence to inhibition by the organophosphorate ester and inactivation by heat.

Measuring Resistance

Once *Boophilus microplus* was proved to be resistant to several active ingredients, various techniques for measuring it were developed, all of them based on trials carried out on two stages: larvae or engorged females.

Techniques Based on Larvae

These are the most wide-spread techniques, their main advantages being the possibility of working with a larger number of elements (therefore on a more complete sampling) and a greater speed in obtaining results.

There are two fundamental types:

a) Those where larvae are submerged in solutions, emulsions or suspension of the acaricide, like those of Whitehead (1961), Shaw (1966), Fiedler (1968) and Grillo Torrado and Gutierrez (1969) modified by Caracostantógolo.

b) Those where larvae are put in constant contact with media impregnated with the acaricide as was developed by Stone and Haydock (1962).

A Immersion Techniques

Whitehead Technique. This was developed in South Africa in the 1950's, a time when the problems of resistance to arsenicals and later chlorides arose. It is interesting to point out that Whitehead mainly used acaricides formulated as wettable powders.

Groups of 100 – 200 larvae, 15 – 30 days old, are dipped in different concentrations of the ixodicide in glass tubes less than 6 cm long by 2 cm in diameter for 1 min in a centrifuge of 32 r.p.m. The larvae are then collected using a copper wire mesh 30 disc, washed with distilled water and, still on the copper wire mesh disc, they are dried by contact with blotting paper, enveloped and sealed.

The trials with each dilution of the acaricide are repeated five times, using a total of 700 to 1000 larvae per concentration. A control treated with water is always included: if the mortality rate of the control group exceeds 5%, then the whole trial is annulled. The larvae are kept for 24 h at 25 °C and relative humidity of 80%, after which the reading takes place: mortality is based on the absence of movement and the results are analyzed by probit analysis method, if necessary determining the lethal concentration (LD_{50}) by extrapolation.

Shaw Technique. This is based on immersion for 10 min between 2 Whatman No 1 filter papers, in groups of about 300, 14 – 21-day-old larvae, at different ixodicide concentrations. Then the groups of 70 – 100 larvae are taken for each concentration and enveloped in filter paper, incubating them at 22 – 28 °C with a relative humidity of 80% for 18 – 24 h. Some authors lengthen this period to 72 h.

Once the live and dead larvae are counted in each envelope (each one corresponds to a different concentration), the results are put down on a special paper denominated "logarithmic-probit" where the ordinate (probit) represents percentage of mortality and the abscissa (double logarithmic scale), the concentrations used, then drawing a regression line that gives the 50 and 99 lethal dosis. The same process is followed for the model susceptible strain, and the result of the division of 50 and 99 lethal doses of both strains gives the corresponding factor of resistance.

This technique was widely used by us in Argentina with more than acceptable results, taking into account that these results are relative compared to those observed in field trials.

Figure 8.5 is a sample of the description of results, using Shaw technique, comparing K strain (coumaphos resistance factor 5 and ethion 2.3 and very sensitive to delnav) in a logarithmic probit paper with the susceptible strain 52.

Fiedler Technique. This immersion technique consists in enveloping 100 – 160 eggs in rice paper envelopes. Three weeks after hatching the larvae are submerged in the respective ixodicides at the desired concentrations for 2 s as rice paper (like that used for tea-bags and other infusions) absorbs the liquid very rapidly, retaining it at a rate of 0.01 ml cm^{-2}. They are then dried at 26 °C with 60% relative humidity, and after 70 min the results are read.

According to Fiedler, this technique has the advantage of avoiding the manipulation of larvae, of the active elements reaching the final trial and of eliminating some other factors of error such as evaluating only some of the treated larvae and also carrying out treatments where the contact of the product with the mite is not as close as through rice paper, which makes it more like a plunge dip (Fiedler 1968).

Grillo Torrado and Gutierrez Technique Modified by Caracostantógolo. In this technique, which has fundamental differences with the original one, 14 – 21-day-old larvae are treated (in Grillo Torrado-Gutierrez technique 3 – 18-day-old larvae were used) at double tickicide dilutions (preferably commercialized formulations) starting at different levels according to the characteristics of each one. As a

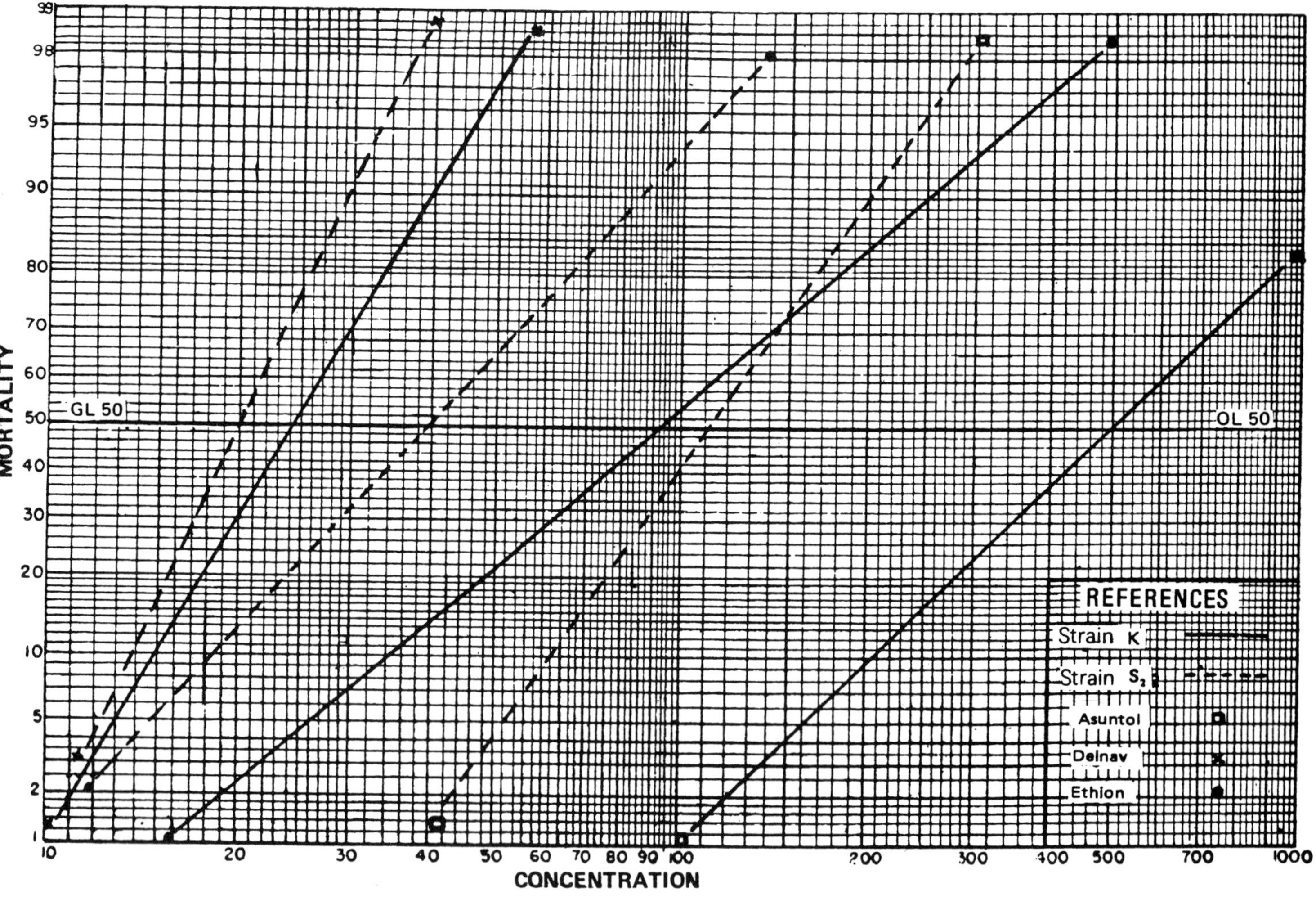
MORTALITY
GL 50
OL 50
REFERENCES
Strain K
Strain S₃
Asuntol
Delnav
Ethion
CONCENTRATION

diluent Triton X100 at 0.0125% in distilled water is used. Time of immersion is 35 to 50 s. The larvae treated in glass tubes are placed on blotting paper and then enveloped in "papel de obra" (a sort of filter paper). This step completed, they are placed in a stove (27.1 °C and 85% relative humidity) for $20-48$ h, depending on the product used.

The results are read by putting down the figures on logarithmic probit paper similar to that used in the Shaw technique, obtaining the factor of resistance (RF) by dividing the 50 and 99 lethal doses of susceptible strain by the resistant one. In the original technique, the Reed and Muench technique was used to find the LD_{50}, which in our opinion is not adequate for this type of trial.

This technique, developed by SELAB, laboratory service of SENASA (Animal Health Service), proved to be regular and sensitive, becoming an important complementary element in the campaign for eradicating ticks which this government office is carrying out.

B Techniques by Contact

Stone and Haydock Technique. This technique was published in 1962. Its main characteristic is the confinement of the samples being tried to a treated surface.

Equal parts of filter paper (Whatman No. 1) of 11 cm diameter are used, where an area of 6 by 4 cm is marked, this being divided into 48 tiny rectangles. In the case of the controls, these papers are impregnated in oil (*Risella oil*) and ixodicides dissolved in the same oil or xylene in the envelopes which will be used to recognize larvae susceptibility to the active principles chosen.

Groups of $100-120$, $7-28$-day-old larvae are placed in envelopes made with this paper folded and sealed with clips. They remain there in a horizontal position for $16-24$ h.

They are then opened and the dead larvae are counted: those which show movement, however slight, are considered to be alive.

The lethal concentration 50 is calculated by means of probit analysis, drawing the regression slope. The correction for mortality in the control group is done by using the Abbot formula.

Technique Recommended by FAO. In this technique, called FAO method No. 7, groups of approximately 100, $7-14$-day-old larvae are treated, placing them in paper envelopes impregnated with olive oil, in which the acaricide at different concentrations is included.

The olive oil is previously sterilized at $105-110$°C for 75 min and when cooled down, 0.02% antioxidant (Isonol) is added. Impregnation of filter paper

Fig. 8.5. *Boophilus microplus can.* In vitro susceptibility trials with 20 Argentine strains. Nuñez et al. (1972)

(Whatman No. 1) is made easier by use of a volatile solvent: chloroform or trichloroethylene.

The enveloped larvae are kept for 24 h at 27 °C with an 80% relative humidity, and the reading is carried out simply by counting the live and dead specimens.

The results are put down on logarithmic probit paper and the 50 and 99 lethal doses are determined by drawing the regression slope.

As in the previous technique, if mortality in the control group exceeds 5%, the whole trial is annulled.

Techniques Based on Engorged Females

In our opinion techniques based on engorged females have one basic defect: the trials must be done with a limited number of adult elements and not with their numerous and varied progeny, which gives a much less realistic picture of the problem, as the sampling is very restricted.

Also, in areas where there is a good control or an eradication plan of periodic treatments, it is difficult to find a sufficient number of engorged females.

In Santa Ana de Livramento, Río Grande do Sul, Brazil, Arteche and colleagues developed a technique for diagnosing susceptibility in engorged females by immersion treatments in vitro and later control of oviposition and mortality.

Stone and Haydock (1962) also adapted their technique, already described, to work with engorged females. In this case they use filter papers of 12×15 cm which are impregnated with the insecticide dissolved in oil. Groups of 25 specimens are treated and placed in ventral decubitus for 48 h. After this period the engorged females are isolated in tubes for oviposition (they are checked after 7 days) and subsequent hatching of eggs (control is done after 28 days).

Techniques based on the use of engorged females are mostly useful to confirm if a chemical compound has tickicide effect or not, but, due to the above-mentioned reasons, its sensitivity is not sufficient to compare the effect of various products.

References

Allan J (1955) Loss of biological efficiency of cattle dipping wash containing benzene hexachloride. Nature (London) 175:1131 – 1133

Arregui LA, Laranja RJ, Arteche CCP (1975) Resistencia do carrapato *Boophilus microplus* (Canestrini, 1888). Determinacao de CIPV 50 in vitro dos garrapaticidas organofosforados usados no Rio Grande do Sul. Frente a teleóginas de una estirpe sencível. Bol Inst Pesqui Vet "Desiderio Finamor" 3:101 – 111

Arteche CCP, Arregui LA, Laranja RJ (1975) Algunos aspectos da resistencia do *Boophilus microplus* (Canestrini, 1888) dos garrapaticidas organofosforados no Rio Grande do Sul. Bol Inst Pesqui Vet "Desiderio Finamor" 4:91 – 99

Arteche CCP, Laranja RJ, Arregui LA (1977) O uso atual dos garrapaticidas arsenicais no Rio Grande do Sul. Bol Inst Pesqui Vet "Desiderio Finamor" 4:13 – 19

Ault CN (1948) Investigaciones sobre las dificultades de combatir la garrapata *Boophilus microplus*. Rev Med Vet 30:174 – 211, 254 – 297

Bennett GF (1975) Oviposition of *Boophilus microplus*, Canestrini (*Acarina, Ixodidae*) III Oviposition pattern of acaricide resistant strain. Acarologia 16:394 – 396

Boero JJ (1953) La resistencia de la garrapata a los clorados. Rev Med Vet 35:169 – 174

Brown ANA (1967) Genetics of insect vectors of disease, chap. 17. Elsevier, Amsterdam, p 505

Chaudhur RP, Naithani RC (1964) Resistance to BHC in the cattle tick *Boophilus microplus* Can. in India. Bull Entomol Res 55:405 – 410

Drummond RO (1977) Resistance in ticks and insects of veterinary importance. Pest. Management and insect. resistance. Academic Press, London New York

Du Toit R, Graf H, Bekker PM (1941) Resistance to arsenic as displayed by the single host blue tick. *Boophilus decoloratus* (Koch) in a localized area of the Union of S. Africa. J S Afr Vet Med Assoc 12:50 – 58

Fiedler OGH (1968) A new biological method for evaluating the efficacy of acaricides against ticks. J S Afr Vet Med Assoc 39 (1):84 – 87

Fluck V, Rufenacht K (1969) Effectiveness of newer phosphorus compounds against resistant ticks of *Boophilus microplus*. Vet Pestic Soc Chem Indust Monogr 33:183 – 193

Grillo Torrado JM, Gutierrez RO (1969) Método para medir la actividad de los acaricidas sobre larvas de garrapatas. Evaluación de la sensibilidad. Rev Invest Agropecu Ser 6: no 4, 135 – 158

Grillo Torrado JM, Gutierrez RO (1970) Fósforo-resistencia de una cepa argentina de garrapata *Boophilus microplus*. Su medición. Rev Med Vet 51:113 – 122

Grillo Torrado JM, Perez Arrieta A (1977) Trasmisión hereditaria de la fósforo-resistencia al Delnav en *Boophilus microplus*. Rev Med Vet 58:309 – 322

Grillo Torrado JM, Gutierrez RO, Perez Arrieta A (1972) El factor de resistencia en larvas de la garrapata *Boophilus microplus* (Can.). Lah a los compuestos organofosforados. Rev Invest Agropecu Ser 4:25 – 35

Harrison IR, Palmer BH, Wilmshurst EC (1973) Chemical control of cattle ticks. Resistance problems. Pestic Sci 4:531 – 542

Jones Davies WJ (1972) Tick control and a history of tick resistance. Rhod Vet J 2:53 – 59

Lee RM, Batham P (1966) The activity and organophosphate inhibition of cholinesterases from susceptible and resistant ticks (Acari). Entomol Exp Appl 9:13 – 24

Matthewson MD, Wilson RG, Hammant CA (1976) The development of resistance to certain organophosphorous and carbamate insecticides by the blue tick *Boophilus decoloratus* (Koch). (Acarina-Ixodidae) in Rhodesia. Bull Entomol Res 66:533 – 560

Morel PC (1967) Résistance de la tique *Boophilus microplus* contre le gaméxane aux Antilles françaises. Rev Elev Med Vet Pays Trop 20:291 – 299

Nolan J, Schnitzerling HJ, Schunter CA (1972) Multiple form of acetylcholinesterase from resistant and susceptible strains of the cattle tick *Boophilus microplus* (Can.). Pestic Biochem Physiol 2:85 – 94

Nolan J, Roulston WT, Wharton RH (1977) Resistance to synthetic pyrethroids in a DDT-resistant strain of *Boophilus microplus*. Pestic Sci 8 (5):484 – 486

Nuñez JL, Pugliese ME, Shaw RD (1972) *Boophilus microplus* Can. Pruebas de susceptibilidad in vitro con veinte cepas argentinas. Rev Med Vet 53:37 – 43

Oliver JA (1967) Cytogenetics of acarines. Genetic of insect vectors of diseases. Capítulo 13. Elsevier, Amsterdam, pp 417 – 439

O'Sullivan PJ, Green PE (1971) New types of organophosphorous-resistant cattle ticks (*Boophilus microplus*). Aust Vet J 47: no 2, 71

Panel de Garrapata (1972) Pracive 1:55 – 56

Reich CI, Grillo Torrado JM, Perez Arrieta A, Zorzopulos J (1978) *Boophilus microplus*: strain differences of the cholinesterase system (personal communication)

Roulston WJ, Wharton RH (1967) Acaricide tests on the Biarra strain of organophosphorous-resistant cattle tick *Boophilus microplus* from Southern Queensland. Aust Vet J 43:129 – 134

Roulston WJ, Schunter CA, Schnitzerling HJ (1966) Metabolism of coumaphos in larvae of the cattle tick *Boophilus microplus*. Aust J Biol Sci 19:619 – 633

Roulston WJ, Schunter CA, Schnitzerling HJ, Wilson JT (1969) Detoxification as a resistance mechanism in a strain of *Boophilus microplus* resistant to organophosphorous and carbamate compounds. Aust J Biol Sci 22(6):1585 – 1589

Schnitzerling HJ, Schunter CA, Roulston WJ, Wilson JT (1974) Characterization of the organophosphorous-resistant Mt. Alford, Gracemere and Silkwood strains of the cattle tick *Boophilus microplus*. Aust J Biol Sci 77:397 – 408

Schunter CA, Smallman (1972) Cholinergic systems in organophosphorous-resistant and susceptible larvae of the cattle tick *Boophilus microplus* (personal communication)

Shaw RD (1966) Culture of an organophosphorous-resistant strain of *Boophilus microplus* (Can.) and an assessment of its resistance spectrum. Bull Entomol Res 56:389 – 405

Shaw RD, Cook M, Carson RE (1968) Developments in the resistant status of the southern cattle tick to organophosphorous and carbamate insecticides. J Econ Entomol 61:1590 – 1594

Stone BF (1962a) The inheritance of DDT resistance in the cattle tick *Boophilus microplus*. Aust J Agric Res 13:984 – 1007

Stone BF (1962b) The inheritance of dieldrin resistance in the cattle tick *Boophilus microplus*. Aust J Agric Res 13:1008 – 1022

Stone BF (1968a) Inheritance of resistance to organophosphorous acaricides in the cattle tick *Boophilus microplus*. Aust J Biol Sci 21:309 – 319

Stone BF (1968b) Brain cholinesterase activity and its inheritance in cattle tick *Boophilus microplus* strains resistant and susceptible to organophosphorous acaricides. Aust J Biol Sci 21:321 – 330

Stone BF (1972) The genetics of resistance by ticks to acaricides. Aust Vet J 48:345 – 350

Stone BF, Haydock KP (1962) A method for measuring the acaricide susceptibility of the cattle tick *Boophilus microplus*. Bull Entomol Res 53:563 – 578

Stone BF, Wilson JT, Youlton NJ (1976a) Linkage and dominance characteristics of genes for resistance to organophosphorous acaricides and allelic inheritance of decreased brain cholinesterase activity in three strains of the cattle tick *Boophilus microplus*. Aust J Biol Sci 29:251 – 263

Stone BF, Nolan J, Schunter CA (1976b) Biochemical genetics of resistance to organophosphorous acaricides in three strains of the cattle tick *Boophilus microplus*. Aust J Biol Sci 29:265 – 279

Vilenberg G (1963) Résistance à l'héxachlorocyclohéxane d'une souche de la tique *Boophilus microplus* (Canestrini) à Madagascar. Essais préliminaires sur sa sensibilité à quelques autres ixodicides. Rev Elev Med Vet Pays Trop 16(2):137 – 146

Wharton RH, Roulston WJ (1970) Resistance of ticks to chemicals. Anu Rev Entomol 15:381 – 404

Whitehead GB (1958) Acaricide resistance in the blue tick *Boophilus decoloratus* (Koch). Part I. Bull Entomol Res 49:661 – 673

Whitehead GB (1961) Investigation of the mechanism of resistance to sodium arsenite in the blue tick *Boophilus decoloratus*. J Insect Physiol 7:177

Whitnall ABM, Thornburn JA, McHardy BM, Whitehead GB, Meerholds F (1952) BHC-resistant tick. Part 1. Bull Entomol Res 43:51 – 65

Wilson RG (1978) Biochemical mechanisms causing tick resistance. J S Afr Vet Assoc 49:49 – 51

Subject Index

J. K. Matsushima

Feeding Beef Cattle

1979. 31 figures, 23 tables. IX, 128 pages
(Advanced Series in Agricultural Sciences,
Volume 7)
ISBN 3-540-09198-X

Contents: Nutrients. – Classification of
Feeds. – Procedures in Ration Formulation. –
Processing Feeds for Beef Cattle. – Systems
of Feeding. – Feed Additives. – Growth
Stimulants.

Written by one of the world's foremost
experts in beef cattle feeding, this timely
book details the nutrient requirements of
cattle as well as the consequences of excessive nutrient feeding. The various feeds available and their relative values in beef cattle
diets are discussed along with the climatic
conditions, types of soil, and the economics
of producing certain feeds. Substitute feeds
are dealt with in depth. The expediency of
different cattle feeding programs is outlined,
and the use and value of feed additives and
growth stimulants are considered.
The author, whose experience spans a period
of thirty years, has had a far-reaching impact
on the growth and development of the
feedlot industry. His up-to-date and compre-
hensive book offers scientists, beef cattlemen
and students all the facts they need to under-
stand beef cattle nutrition and to plan feeding
programs suited to individual needs.

Springer-Verlag
Berlin
Heidelberg
New York
Tokyo

J. R. Parks

A Theory of Feeding and Growth of Animals

1982. 123 figures. XVI, 322 pages. (Advanced Series in Agricultural Sciences, Volume 11)
ISBN 3-540-11122-0

Contents: Introduction. – Ad Libitum Feeding and Growth Functions. – A Stochastic Model of Animal Growth. – Treatment of Ad Libitum Feeding and Growth Data. – The Geometry of Ad Libitum Growth Curves. – Growth Response to Controlled Feeding. – The Theory. – A General Euclidean Vector Representation of Mixtures. – The Effects of Diet Composition on the Growth Parameters. – The Growth Parameters and the Genetics of Growth and Feeding. – Energy, Feeding, and Growth. – Appendices. – Glossary of Mathematical Symbols. – Glossary of Words and Phrases. – Subject Index.

The subject of this book is the search for the deterministic elements in animal feeding and growth patterns that could form the basis of a testable theory. The first six chapters are devoted to mathematical study of past growth functions as related to numerous sets of experimental growth and feeding data for animals on various diets and under various feeding regimes. Chapter seven integrates these studies into a mathematical theory of feeding and growth and illustrates its use in a long-term experiment on two genotypes of chickens. The theory was sufficiently robust to be used further in studies on the diets and the nutrition of other growing animals, the results of which are reported in chapters eight and nine. The next chapter discusses the implications of the theory in the genetic experimental work on bending the growth curves of mice and chickens by selection techniques and in the economics of intensive animal production. Chapter eleven finally relates the theory of the energy balance all animals as open systems must obey, making this book a valuable guide to laying a scientific foundation for any undertaking in animal management and production technology.

Springer-Verlag
Berlin
Heidelberg
New York
Tokyo